THE MODEL RAILROADER'S GUIDE TO
LOGGING RAILROADS

MATT COLEMAN

KALMBACH
BOOKS

About the author

Matt Coleman's lifelong interest in trains was kindled by his father, a steam fan who took his son trackside to watch locomotives before Matt could walk. Matt's varied professional career has encompassed 25 years as an engineer, scientist, and technical and scientific editor in the forest products industry, including a 10-year stint as the editor and publisher of the *Tappi Journal,* a noted peer-reviewed scientific publication serving the forestry sciences, paper, and packaging industries.

Matt lists his primary modeling interests as logging and forestry railroads, both foreign and domestic, in a variety of scales and gauges. He also holds a deep passion for the Denver & South Park narrow gauge lines in Colorado, the Sandy River in Maine, and an equally incompatible love for the Atchison, Topeka & Santa Fe, especially its gas-electric operations in the Southwest and West. Matt is married to Robin, a patient, tolerant, and loving wife, and between them they have four children.

Printed in China

23 22 21 20 19 18   8 9 10 11 12 13 14 15 16 17

Visit our Web site at KalmbachBooks.com.
Secure online ordering available.

Unless noted, photos were taken by the author

Publisher's Cataloging-in-Publication Data
(Prepared by The Donohue Group, Inc.)

Coleman, Matthew.
   The model railroader's guide to logging railroads / Matt Coleman

   p. : ill. ; cm.

   ISBN: 978-0-89024-702-0

1. Logging railroads--Models. 2. Logging railroads--History. 3. Logging--History. 4. Lumbering--History. I. Title. II. Title: Logging railroads

TF197 .C65 2007
625.1/9

CONTENTS

Introduction

Writing a single book about logging and logging railroads is akin to trying to pack for a two-year journey around the world in a single suitcase. No matter how careful the planning, somewhere on the slopes of Mount Kilimanjaro or some other remote spot you will find that something you truly needed has been left out and something that is utterly unnecessary has been packed. And so it is with this book.

As I started writing I knew that, because of the broad nature of the topic, the book would be incomplete in some areas. I have spent the better part of my civilian professional career in the forest, paper, and related scientific fields, and have devoted much of my hobby time researching and modeling logging railroads. I hope that this book provides a solid overall description of the subject and inspires you to study more and dig deeper into specific areas of interest regarding logging operations and railroads.

Over the years, I have become an avid model railroader. Although I love prototype railroads and have chased trains on four continents, my true affection always returns to scale models of trains. Within that focus, my interest in models of railroad equipment that hauled logs and wood products ranks near the top of the list. As a result, this book covers the wonderful world of prototype logging, but always with a focus of how it could apply to a model railroad.

My early years of exposure to trains were the result of being born in 1950 in the middle of the post-war golden age of model railroading. By age 3, I had a Marx train set and by age 6, I had been given a small HO display layout with a Penn Line midget diesel. I literally wore the wheels off of it due to constant running. I was also blessed with a younger brother who also loved trains, as well as a father who, while not a model railroader, was both supportive and a devotee of all things steam-powered. This exposed me to, and helped me to see, many aspects of the real world that might otherwise have escaped the eyes and ears of even the most avid 10-year-old model railroader.

In addition to the occasional cab ride or visit to a steam locomotive, my parents willingly stopped the car to let me peer into canyons and valleys to look at old, graying trestles of logging railroads across the country, from New Mexico up to Oregon and British Columbia, to the pines of Michigan and the South, and to the laurel-covered hills of Pennsylvania and the Appalachians. I visited museums, saw prototype logging lines in action (albeit with diesels hauling trains by the 1960s), and came to see the fading tracks of the logging line into the woods as the true "narrow path" that I longed to follow with my modeling.

But all of this exposure would have been almost meaningless were it not for one other aspect of the hobby that is almost as important to me as prototype equipment and models. That is the publications, photographs, and other historical remnants that have made it possible for those of us in later generations to learn about and accurately model railroads of all types, in particular logging railroads. I have a collection of *Model Railroader* magazines that, although incomplete, goes back to volume 1, issue 1 of 1934 (although a reprint), and I must confess that my years of reading *Model Railroader* have indelibly shaped my view of model railroading. There are many fine magazines and editors in this hobby, but my most treasured issues of all publications seem to be from the Kalmbach presses. I even have a specific issue of a magazine that was a turning point in my view of the hobby, one that I have treasured: the September 1961 issue of *Model Trains*, which was a key part of my interest in tall timber and geared locomotives.

My goal in writing this book is to try to pass information on logging to a new generation of modelers who were never able to see what I was able to see in my early years. This generation, with more and better models to work with, has the potential to create far better logging-related model railroads than could any of us in the past. To them, and to all who helped preserve the history of logging railroads and their equipment, I dedicate this book.

– *Matt Coleman*

Why model a logging railroad?

For many, the answer is just "I like the equipment" or "I like geared locomotives," but for others, logging lines hold a romance and an interest that demand to be incorporated into an existing model railroad. No matter what the reason, logging railroads are perennially popular with modelers and have been for more than half a century.

Let's start with a look at some of the characteristics of pure logging railroads to help us understand what we are trying to model. For starters, a prototype logging railroad was almost always strictly an industrial line. Logging trains hauled logs from the cutting or loading site to the sawmill and returned with the empty cars. These railroads were conveyors on rails, operating unit trains long before Class I railroads began doing so in the 1960s.

An early two-truck Shay eases a log train down the mountain toward the mill. The giant fir logs are riding on wood-beam disconnect trucks. *Darius Kinsey, Library of Congress*

5

Larger logging lines, which were usually part of a major corporate enterprise such as Weyerhaeuser, Georgia-Pacific, or International Paper, featured periodic movements of equipment such as loaders or skidders. Occasionally the entire logging operation – bunkhouses, cook-house, and all – would be loaded onto railcars and hauled gingerly and slowly to a new site.

The purpose of these moves was purely economic – the company wanted to keep the loggers as close to the timber cutting face as possible to cut down on transit time.

Trains did not operate in the same manner as on common carrier railroads. The ideal, from the logging company perspective, was to have every train loaded to maximum capacity from the cutting area and have every day be the same as the day before with no interruptions to traffic. The purpose of the logging railroad was to keep a steady flow of logs coming to the millpond or cold-deck (unloading area) so the sawmill never had to shut down.

Smaller logging operations and earlier (pre-1900) lines had much more of a common-carrier rail-road flair to them, even if they were not chartered as such. Less-

strict regulations meant that a logging railroad could be built and run at a lower cost than a common carrier. However, convenience as well as simple necessity made many logging lines the only way in or out of some remote areas, and those who lived in those areas used logging trains for transportation as well as for shipping and receiving goods. This gives us additional modeling opportunities.

Logging lines that were chartered as common carriers were usually incorporated that way for tax purposes or to protect their territory from encroachment by larger common-carrier railroads. Some logging railroads were made common carriers in the vain hope of being purchased by a larger railroad, with owners hoping to make money from the timber along the line and reap a second profit from the sale of the railroad. Few of these schemes were successful, but they made for some interesting and colorful operations.

Some of the best known of these thinly veiled loggers were the Sandy River & Rangeley Lakes two-foot-gauge lines in Maine. While described in books as mostly a common carrier line, the Sandy River, Phillips & Rangeley,

A log is carried up the green chain from the pond to the sawmill at a British Columbia lumber mill in the 1980s.

Kennebec & Dead River, and Eustis lines were primarily built to haul logs and pulpwood from the forests to paper mills and sawmills in Maine. More than 90 percent of the aggregate tonnage of the Sandy River and predecessor lines was cut timber; the rest was mostly finished wood products. Yet most of us remember these lines for their fast passenger trains to resorts, tiny parlor cars, and magnificent examples of

A landing donkey (steam-powered winch) pulls logs toward the water. What was a remote wilderness area at the turn of the 20th century is now a suburban Seattle community. *Darius Kinsey, Library of Congress*

A small rod locomotive brings in a train of modestly sized logs (for the Northwest) as family members or friends ride on the pilot. *Library of Congress*

Baldwin's most beautiful small locomotives.

Many logging railroads operated in beautiful, picturesque areas of the country, including the majestic coast range mountains of Oregon and Washington, the wooded coves and hollows of the West Virginia Appalachians, the rugged, rocky terrain of the Sierra Nevadas in California, and the coastal redwood area of northern California.

Equipment and variety

Logging railroads ran some of the most fascinating – and modelable – equipment ever operated. Geared steam locomotives are so slow that even on a small layout they take significant time to cover the run.

Most logging railroads were temporary, built to last for just a cutting season or two. After that, the rails would be removed, the equipment hauled away, and the track, like a tinplate layout on a living room floor, would be re-laid to serve again in a different location.

Because of this, logging line roadbeds were often twisting, writhing paths that went along hillsides and canyon walls, up steep grades, and over spidery timber trestles.

Logging lines operated everywhere there were forests. Few states had no officially chartered logging railroads. (Connecticut, Illinois, Indiana, Kansas, Nebraska, North Dakota, Ohio, Rhode Island, and Utah had no logging lines.) Some of these states did feature logging by rail, but it was on common-carrier railroads.

Even today, on such lines as the Georgia Northeastern (which operates a portion of the old Louisville & Nashville "Hook and Eye" route through the Georgia and Tennessee mountains), carloads of logs regularly trail well-kept first-generation EMD diesels.

The most successful logging lines were those in the Pacific

On early logging lines, a locomotive hauled a string of cars directly to an A-frame unloader, dropping a string of multi-ton logs into the water in a matter of minutes. This early 1900s scene is on California's Northern Redwood Lumber Co. *Two photos: Edw. R. Kirkwood*

Northwest and other western mountainous areas. Here, truck competition was limited, and a vast supply of first- and second-growth timber provided a steady stream of logs to reloading points along established "main lines" in the woods.

The Weyerhaeuser lines on the lower Columbia River, the West Side Lumber Co., the Georgia-Pacific operations in the Oregon coastal range, and the Simpson Timber Co. all lasted into the 1960s and '70s as log haulers. These survivors depended on reload points, where big diesel trucks brought logs down from cutting areas for reloading onto railcars. A locomotive with 20 to

25 log cars took over for the run to the mill.

Logging's intrigue for modelers has remained steady, despite a steady decline in the number of operating logging railroads. Perhaps this is because logging lines allow more freedom of modeling and yet are more challenging in terms of equipment and scenery. Modeling one tree well is a task, but modeling the hundreds needed for even a simple logging railroad can be a significant challenge.

I think the real reason logging railroads remain popular is the romance of the industry itself. Logging was one of the first – and most important – industries in the

A diminutive Southern Pacific 2-8-0 pulls a log train on the Cloudcroft branch near Alamogordo, N.M., in an example of logs being hauled by a common-carrier railroad. This view is from 1946, three years before the line was abandoned. *Henry Garcia*

Photos like this one, which illustrate the enormous size of Western trees such as the Douglas fir, were used in the late 1800s to encourage Eastern investors to back Western logging operations. *Library of Congress*

United States. Hundreds of books and articles have been published, documenting almost every aspect of logging and sawmilling, as well as the railroads that served those parts of the industry.

Early U.S. logging

Native tribal cultures of North America included many references to the forest in their stories and oral histories, and it was clear that they understood what the forest meant to their way of life.

For European settlers, the forest was both a treasure and an impediment. Forests were in the way of planting the crops that were necessary for survival, but within a few years, the trees themselves were the only crop that mattered in the New England colonies. The view of trees as a cheap resource for building homes and for use as fuel, especially in the coastal and riverine areas of Maine, changed once the English Crown realized that the large, plentiful trees held strategic value as masts for Royal Navy ships. In the late 1600s, an edict went out declaring that any tree

over 24 inches in diameter was Crown property and could not be harvested. The increasingly restless colonialists began to subvert this order, harvesting the giant pines as they saw fit and covering their actions by simply making sure that any board cut from them was always less than 24 inches. Royal Navy and Army inspectors were eventually sent to protect the trees, adding to the conflict between England and the colonists.

Pre-industrial operations

The first settlers in any area began cutting trees to create primitive shelters. These are often called "log cabins," although that term actually came into use long after that type of dwelling had fallen into disuse. Once the cabin had been built, trees were cut and the land cleared (sometimes with the stumps left in the ground) for farm fields.

The ingenuity and skills of the early carpenters then came into play. Through the colonial era, logs were harvested by double-bladed axe and cut into boards either by splitting with a wedge and froe or by sawing with a double-handled saw using a raised

The three-foot-gauge West Side Lumber Co., in Tuolumne, Calif., is a favorite of modelers. Here two-truck Shay No. 10 rounds a curve in October 1959. *Robert Field*

platform and a pit. The amount of labor needed to create a single hand-cut board provides an idea of the wealth needed to build a house with siding instead of split or plain logs. The saw pit was an institution in most colonial villages for most of the 1700s, and not until the larger use of water power in the early 1800s did the mechanization of sawmills begin in earnest.

The difficulties of cutting and moving a large log were almost overwhelming. It's no accident that most logging first took place along rivers, where logs could be floated down to the mill during the spring thaw. This was a risky business and a major limitation to the use of sawed timber.

If you weren't near a river, the only way to move a log was with animal power. Anyone who has ever cut down a tree, even a small one, understands how much a tree weighs and how difficult it is to move. Imagine the forces required to move a 40-foot-tall, 30-inch-diameter oak. A team of four oxen would do, but only if the path was level and the distance was less than a half mile or so. Then, when you got the log to the saw pit, moving it into position would take all the oxen and several men. Cutting the log into 1- or 1½-inch boards for furniture could take a week or more of the efforts of two strong men.

Steam was the savior of the timber business. In England,

The steam donkey, or skidder, a portable machine used for moving and loading logs, has a vertical boiler with one or more winches. They were common by 1900. *Library of Congress*

steam engines were first developed to pump water from the coal mines of Newcastle, but steam soon powered railway locomotives and other machinery. By 1824, at least three steam sawmills were operating in Pennsylvania, along with others elsewhere around the country.

By 1900, the U.S. had more than 180,000 miles of common-carrier railroads. All used wood ties, and at about 3,000 ties per mile, you can sense the scope of the involvement of railroads with the timber industry. Add to this trestles, structures, culverts, telegraph poles, and the wood used in the rolling stock, and it's little wonder that the country's

total harvest of wood peaked between 1890 and 1910, which also corresponds with the peak sales years for Shay geared locomotives and logging locomotives of all kinds.

Even today, timber harvested in the forests of the U.S. and Canada is the primary framing material for more than 80 percent of the structures built in the U.S. Most of that wood travels by rail at some point between the forest and the final point of use.

The timber industry today
Logging is a huge industry with many facets. I have seen logging operations, sawmills, and paper mills on four continents and in 17

This late-19th-century view shows logs taken from a single California redwood tree in a train pulled by a 2-4-2T. *Library of Congress*

This early lumber camp and mill at Concord, N.H., is typical of the appearance of Eastern mill complexes through the 1920s. Note the chain from the log pond to the mill and the buildings built of logs. *Library of Congress*

Camp workers pose with several cut logs on the skidway, used for sliding logs to the loading area. *Library of Congress*

countries, and yet I can truthfully say that I am not an expert on any facet of this industry. I belong to several on-line discussion groups, and there isn't a day that I don't learn something new from one of the groups. (My favorite is the 4L Logging Group on Yahoo (http://groups.yahoo.com/group/4L/.) As a result, I have come to see that there are no logging experts in our hobby, but rather there are some people who know a great deal about specific companies, locations, or equipment – but no one knows it all. Collectively, that group has a far larger body of knowledge than any single book or museum.

Modeling limitations

Logging railroads have their limitations in terms of operational possibilities and variety. Each is a highly specialized industrial railroad hauling one specific product (logs, pulpwood, or wood chips) from one place to another. There are no passenger trains except perhaps a morning and evening logger car attached to a train to bring crews into and out of the woods (or, in later years, a gas-mechanical passenger vehicle that carried the loggers).

Logging trains rarely engaged in backing movements, especially when using disconnect trucks. Until the last years of logging railroads, when all the cars were steel-framed with air brakes, backing movements with heavily loaded cars were avoided at all costs. We will cover more on logging car operational limitations in later chapters, but for now realize that to accurately replicate the actions of a logging operation requires a different (and to some an enjoyable) design of layout and operations.

To make a logging railroad look realistic, ordinary commercial track is not viable. In HO, code 70 is the absolute largest rail you could use, and then only on the major trunks of a reload line where the large western logging Mallet steam locomotives from Baldwin roamed. Better-looking layouts use code 55 or 60 rail to more accurately replicate the railroads that ran into the woods. Track is a scale model too and should be built to scale just like your cars, buildings, and locomotives.

Logging railroads are, to me, among the most enjoyable and challenging subsets of model railroading. You can get started with a small locomotive and a few flat cars, but to really capture the beauty and uniqueness of a logging railroad, more effort and more research is required than in many other aspects of the hobby.

Several former logging railroads have been revived as tourist lines. This 1980s view is on the Westside & Cherry Valley, which operated for a time on the original West Side Lumber Co. yard and sawmill area in Tuolumne, Calif.

Types of logging railroads

Weyerhaeuser's line out of Klamath Falls was one of the last log haulers in Oregon and put on an impressive show. The company's rebuilt GE U25Bs are shown here in the mid-1970s cresting a hill 10 miles out of Klamath Falls on their way to the mill at Bly.

Not all logging railroads are created equal. It can be difficult to categorize the many types of logging lines because so many variables were involved. First on the list are the so-called "pure" logging lines. These self-contained, geared-steam-locomotive-powered operations are the ones that generally come to mind when discussing logging railroads. Several other types of logging railroads also operated. There were long-haul lines that included several branches, reload lines that depended on trucks to haul logs from the cutting area to the railcar-loading area, and corporation-owned branch or short lines that may or may not have hauled other freight as well. Combined with equipment and practices that evolved significantly over the years, it's difficult to say any one line was a "typical" logging railroad.

This steam skidder has its own tower. It's being used to load a log car. *Weyerhaeuser Historical Collection*

This Pacific Northwest scene shows a small Climax geared locomotive at the loading area, with a forest of tall trees dwarfing the train. *Darius Kinsey, Library of Congress*

The "pure" logging line

This category represents the largest number of forest-related railroads, which could be found over most of the United States and Canada. The simple definition of a pure logging line is that it ran from where the trees were cut to where they were sawn into timber. The railroad's sole purpose was to bring a steady flow of cut and de-limbed tree trunks to a sawmill, so it had only the equipment necessary to do this function.

No direct connection with other railroads was needed (although they often had a connection), and it was this self-contained aspect that allowed some of the most fascinating and colorful narrow-gauge logging lines to survive late into the diesel era. For example, as late as 1960, California's West Side Lumber Co. was bringing Sierra pine to the sawmill behind beautiful little 60-ton, three-truck Shays – while at the same time second-generation diesels were rolling off production lines at Alco, EMD, and GE.

The dedicated logging railroad was the primary way that the

This turn-of-the-century log train on the flatlands of western Washington (powered by an older 4-4-0) shows that early logging railroads weren't always side-of-the-mountain enterprises. *Darius Kinsey, Library of Congress*

A small two-truck Shay has just arrived with its redwood tree load at a northern California mill. *Library of Congress*

majority of harvested timber traveled from the woods to the mill from 1880 through the late 1920s. This is the era that largely shaped our understanding and view of logging railroads, and it is the primary source of inspiration for many model logging lines.

These railroads varied with the era and the region in which they operated. The size and type of timber were responsible for regional variations. For example, hardwood trees in the East and Northeast were more modest in size than the giant Douglas fir and redwood trees of the Cascade Range of Oregon, Washington, and California, which required massive logging machinery and railroad equipment.

The Pacific region became one of the best-known subcultures within the logging industry, with many companies building machinery designed specifically to serve the type of logging in that area. John Labbe's famous book *Steam in the Woods* provides an excellent portrayal of this region during the late logging era.

This region spawned the Willamette geared locomotive, built in Portland, Ore., which forced locomotive giant Lima to offer the specialized Pacific Coast Shay. Heisler and Climax also developed large three-truck versions of their locomotives, employing superheaters, piston valves, and improved valve gear to meet the efficiency needs of giant corporations such as Weyerhaeuser, Georgia-Pacific, Boise Cascade, and Louisiana-Pacific.

Dedicated logging lines were the most prolific users of the geared locomotives that most modelers think of as logging engines: Shay, Climax, and Heisler. However, these geared locomotives were actually in the minority, as most logging railroads relied on conventional (rod) locomotives, often former common-carrier locomotives obtained second-hand.

A steam skidder/yarder pulls five-foot-diameter logs into position at the loading deck in this 1902 scene. *Library of Congress*

The crew at a camp in Michigan pauses for a photo portrait after dinner. Note the clothing typical of the late 1800s and early 1900s. *Library of Congress*

There was a distinct type of logging line that employed geared locomotives, and even on those lines, rod locomotives often handled the longer hauls. This has as much to do with speed as with the ability to take curves or grades. With Shays limited to a 10 to 12 mile-per-hour top speed (and they often ran slower than that), it would be hard to imagine using them on a turn that was longer than 20 miles, since it would take most of the day (and use a lot of fuel and water) just to make a single trip from the loading site to the sawmill.

Long-haul logging line

As logging railroads grew, notably in the Pacific Northwest, the amount of available timber and the costs associated with the old method of moving the sawmill when the land was cut over created a unique kind of logging line. Long-haul lines featured temporary branches worked by geared locomotives that brought loaded log cars to junctions deep in the mountain forests. From there, larger rod locomotives brought cars down to the mills along the lower Columbia River and along Puget Sound.

This unique Shay locomotive, shown around 1900, operated on inventor Ephraim Shay's own logging line in Harbor Springs, Mich. Ephraim Shay ran the locomotive as a test-bed for geared-locomotive development. *Library of Congress*

The St. Maries River Railroad is owned by a lumber company but also serves other customers as a common carrier. The Idaho line was once part of the Milwaukee Road.

These operations were similar to the Class I pattern of branch and trunk lines, but with a key difference: They had no on-line industries, making every train both an express and a unit train. With no local switching, the focus was on safely moving carloads of timber to the mill.

Railroads owned by large corporations could afford equipment designed specifically for their own needs. Examples include amazing low-drivered Mallet locomotives, from 2-4-4-2s up through a giant 2-8-8-2 with 51-inch drivers, which worked one Weyerhaeuser line. This 2-8-8-2, according to Baldwin promotional literature of the time, could haul enough logs on one trip to keep a three-saw mill going for 24 to 36 hours.

Most models of geared and rod logging locomotives that have been produced in brass are patterned after locomotives of these giant Western logging companies. Chapter 5 has more details on how these operations worked.

The reload line

In the 1930s, logging operations began to evolve with advances in technology. Internal combustion came to the woods, not on the rails but in trucks. Although early trucks were fairly light and could haul only one large log, or two or three small logs, they opened up many small logging areas that until that time had been too far from the tracks to be economically viable.

Another internal-combustion breakthrough was the powered chain saw. They allowed a crew of four or five men to fell as many trees in a single morning as could have been cut in a day by 20 men using two-man "misery whips," as the double-handled bucksaws were known.

Spurred on by the development of more-powerful engines during World War II, logging trucks increased in size and began taking over the job of moving logs from the cutting area. In many locations, the primitive roads didn't allow log trucks to travel all the way to the mill. Thus, on larger lines the trucks came to a reload point where big Willamette and Lidgerwood skidders and loaders transferred logs from trucks to railcars. In some cases, a storage area for the logs (commonly called a cold deck or log deck) was established at the railhead, with entire trains loaded at once before being shipped out.

These semi-permanent reload areas married the flexibility of the diesel truck in bringing logs from the cutting sites with the efficiency of the railroad in hauling entire trainloads of logs to the mill on existing, usually paid-for, rail lines.

Few growing towns wanted their city streets clogged with giant trucks loaded with logs, and the logging railroads did a good job keeping the trucks in the woods and the logs on the rails until the 1950s.

Some of these reload lines lasted into the 1990s in the West and upper Midwest. Oregon's Coos Bay Lumber bought dynamic-brake-equipped EMD SW1200s to replace tiny-drivered Baldwin 2-8-2s. Big Baldwin diesel switchers replaced Baldwin 2-8-2s and 2-6-2s in Arizona and California. In Wisconsin and

After World War II, trucks took over for log skidders in bringing logs from the cutting area to the spar-pole loader, where the logs were transferred to railcars. This is MacMillian & Bloedel's operation in British Columbia. *Clinton Jones, Jr.*

Michigan's Upper Peninsula, even more-modern diesels moved logs from reload points to mills.

In later years, reload lines often served fully integrated forest product mills. These facilities, especially in the South and upper Midwest, featured a paper mill, sawmill, a plywood/chipboard plant, and a wood chemicals facility, all at a single location. These sites first began to appear in the

late 1950s and blossomed in the 1970s. They can be as complex as a steel mill, with more variety in operations than a pure logging or simple reload line.

Junctions with main lines

For modelers who have existing layouts and want to operate a Shay or Heisler alongside their USRA Mikados and gas-electric doodlebugs, another aspect of

Diesels replaced steam locomotives in the latter years of Southwest Forest Industries' operations near Flagstaff, Ariz. This first-generation Baldwin switcher is pulling log loads on heavy-duty steel flatcars. *Eric Lundberg*

Argent Lumber Co. No. 3 – still a wood-burner as it approached the end of its service life in the mid-1950s – waits at the reload area as logs are transferred from trucks to log bunks for the trip to the mill. Note the link-and-pin slot in the pilot beam. *Jim Shaughnessy*

logging lines can be modeled: the junction where the sawmill (finished-product) line meets the common-carrier main line. Two examples are the St. Regis paper line junction with the Spokane, Portland & Seattle near Klickitat, Wash., along the Columbia River, and the Feather River Railway, owned by Georgia-Pacific, in California's Feather River Canyon. Both operated with steam well into the late 1950s, with Shays bringing boxcar loads of finished timber from the sawmill to the common carrier junction and taking empty cars back up to the mill.

All that's needed to model such a scene is a small yard with a couple of double-ended sidings where the mainline railroad can interchange cars with the logging line. Although you won't have the variety of cars (no log cars were interchanged), you can still have a three-truck Shay out there next to 4-8-4s and 4-6-6-4s – and even Alco FA and RS-3 diesels if you model the SP&S!

The logging branch

Contrary to what many model railroaders believe (and portray), it was quite rare for log loads to travel on common-carrier railroads. Into the early 1900s, some railroads operated logging branches off of their main lines. These branches went through logging areas, with logs loaded onto flatcars and taken, like any other load, to a designated sawmill.

These operations ended around 1910, as new safety regulations were developed regarding logs and other loose loads. Also, common carriers didn't want to run slow-moving, prone-to-derailment log trains on the same tracks that they were operating passenger and fast freight trains.

By the teens, few new railroads were being built. As a result, it was rare by then for common-carrier branch lines to pass through unlogged areas, meaning fewer opportunities for hauling logs.

There are quite a few logging branches still operating today, but they are almost exclusively used for hauling loads of pulpwood and wood chips.

Company-owned branches

This company-owned, or "captive," branch is almost indistinguishable from a common carrier except that it often has only one customer (or two or more related customers).

Starting around 1900, the larger companies in the forest products businesses began to emulate the mining and steel industries, chartering their own captive common-carrier railroads to serve their facilities. This had economic roots, as the percentage of shipping charges applied to an originating road were much higher than for a railroad that merely moved a boxcar of wood between interchange points. The arrangement also afforded a certain amount of legal protection in the case of accidents and other damage to private individuals. After all, a small short line wouldn't have pockets as deep as a large timber or paper company.

Such lines still exist, and had a resurgence of sorts in the 1970s and '80s as Class I railroads went through rounds of mergers that led to the railroad system we have today. As the large railroads attempted to cut duplicate main lines, some major industries decided to protect their rail service by buying a short line to connect their facilities to the outside world.

Company-owned railroads, especially those that serve large integrated forest-product facilities, have all the characteristics of common carriers, including a wide variety of equipment and heavy traffic. Locomotives are often a bit older and interesting in their own right, but they also have wood chip cars, modern log

The paper industry, although part of the forest products industry, typically has separate, large plants. This is the former Boise-Cascade mill in Wallula, Wash., in 1978.

Paper mills are huge operations, with long, low buildings and separate storage areas for wood chips, logs, and other raw materials. *Matt Coleman collection*

South Carolina's three-foot-gauge Argent Lumber Co. was an example of a pure logging line that became a reload line in its later days. It was abandoned in the late 1950s when the pines were gone. *Jim Shaughnessy*

cars, boxcars, various tank cars, and covered hoppers, as well as flat cars and center beam cars.

Just watching switching activities at the yard of a big facility like the International Paper complex at Cantonment, Fla., or the Georgia-Pacific complex at Crossett, Ark., is inspiring for the constant movement of cars and locomotives. Like a giant switching layout, the movements are slow and deliberate, with loaded cars swaying over well-used track, and the skyline of the giant mill, logs on the cold deck, and wood chip piles as large as sand dunes for a background.

Narrowing the choices

What does all of that mean for us modelers? First, it means that we can't have everything on a model railroad. We all have to make decisions based on our available space, how much we can afford, the era and prototype or locale we plan to model, and our general modeling interests. Most importantly, we need to determine the critical elements most important to us in a model railroad.

For me, having seen Shays in action as a child and in museum settings as an adult, any logging model railroad I build would need to have Shay locomotives snaking through S-curves along the side of a wooded hill, as these are the most vivid images I have of logging. I would want some sort of continuous operation, with a setting perhaps at an intermediate passing siding in the woods where a train of empties coming upgrade to the loading area meets the down-bound loaded train of logs for the mill. The era would be late enough to have three-truck Shays and skeleton log cars with air brakes.

Your list of priorities will differ. Even those with similar interests will come up with different solutions. For example, I have visited four different model railroads based on the West Side Lumber Company. Each one focused on a different aspect of the railroad, and none was even remotely close to being the same.

Given that there were logging railroads in almost every area of the country, and that there is a tremendous variety of model equipment on the market (at least in HO scale), you can come close to capturing at least an approximation of almost any logging railroad that existed. Many modelers will gravitate to the more spectacular or better-known lines, but there is something profoundly

It took experience and good judgment to estimate the amount of lumber that could be obtained from a pile of uncut logs. *Library of Congress*

The earliest colonial-era mill operations involved hand-sawing the logs – a labor-intensive process. *Library of Congress*

interesting in researching and modeling the obscure lines that ran close to home. Consider this as you plan your layout.

Most modelers start a layout with equipment they already own or like, then develop a concept to match. This is especially true in this golden age of model railroading in which manufacturers bring out new, well-detailed models on a regular basis.

One recent development has been the explosive growth of On2½, also known as On30 (O scale narrow gauge, 30 inches between the rails), especially in the area of logging. The Bachmann On30 Shay, although based on a lone prototype (the mill switcher for the Michigan-California Lumber Co. in Camino, Calif.), is such a superb model that it has spawned a variety of aftermarket accessories to modify its era and appearance. This locomotive has enabled modelers to build (or add) a narrow-gauge logging operation at much less cost and difficulty than would have been possible several years ago. Indeed, a small On30 layout could be built, Shay and all, for less than the price of one imported brass Shay in On3.

Matching era and railroad

The easiest type of layout to design and build in terms of equipment availability would be a pure logging line set between 1925 and 1950 or a modern forest-products industry short line in HO scale. There are hundreds of kits and ready-to-run products for these types of layouts. Building a layout in N, O, O narrow gauge, S, Sn3, or large scale requires more effort and is a bit more restrictive in terms of available equipment, but is still quite possible.

If you have an existing layout and want to add a forest-related railroad, look at the obvious before you run out and buy a shiny new Shay or Climax for that

The Valdosta Southern was a captive short line that served a single customer: the paper mill at Valdosta, Ga. The dynamic brakes (the hood bulge in front of the cab) and hood-mounted air tanks give the SW1200 a logging-railroad appearance.

new branch line. If you already model the modern era and have a substantial investment in AC4400 diesels equipped with DCC and sound, adding a backwoodsy timber line powered by a Class A Climax steam locomotive won't necessarily work. It might have some initial appeal, but you'll eventually find the anachronism bothersome, and it will detract from the rest of the layout.

It's more believable to add a steam-powered logging line to an existing steam or steam-to-diesel transition-era layout set in the mid-1950s or earlier. Although a

few Shays still operated in the 1960s, they were few and far between, and only one or two were still in the woods. For the most part, steam logging had ceased by 1961. If you have a modern layout but have fallen in love with geared steam locos, consider building a smaller, separate layout set in the glory days of logging.

You might also consider trying another scale or gauge. It's likely that more Bachmann On30 Shays were purchased by HO and N scale fans that just couldn't pass them up than were bought by diehard O scale narrow gauge fans.

The Corinth & Counce was a common carrier but owned by a paper mill. The mill also operated a sawmill into the 1980s, to which the railroad delivered logs on modern all-steel cars.

Also, the logging world was not populated by geared locomotives alone. The vast majority of logging was done with rod locomotives, many of which were former mainline locomotives that had passed their economic prime. George Abdill's wonderful books on western railroads have many shots of graceful old 4-4-0s and small 2-6-0s bumped from the Southern Pacific or Santa Fe still earning their keep moving cars around the yards of sawmills from Texas to British Columbia.

Remember that rod locomotives built for logging had very small drivers, typically 48 inches in diameter or less – something not found on most ready-to-run models in HO, N, or O. In On30 it would be hard to beat either the Bachmann Porter 0-4-0T or, for On3, the Grandt Line 10- and 15-ton Porter kits, which are some of the finest and most detailed locomotives in any scale. For a small operation – like many in the East and South – one or two Grandt Line Porters and some Keystone On3 skeleton log cars would be the perfect entry into the world of intensely detailed scenes so typical of the larger scales.

You can also consider the possibility of adding a sawmill or paper mill to your layout, with a captive branch line to serve it. Depending on the era, you could power the line with anything from steam locomotives through second generation die-

sels. One fascinating era was the 1990s, when short lines were still running first-generation Alco, Baldwin, and EMD switchers.

The diesel-era reload line falls into this same category, with diesel switchers hauling log trains in the Pacific Northwest, upper Midwest (Wisconsin and Minnesota), and western Canada.

The mill switcher

One final layout consideration is focusing on the industry itself. A model of a large paper mill complex, with a sawmill, a saw-timber log yard, wood chip and pulp-wood storage, chemical recovery plant, and lumber-drying kilns could take up half a good-sized basement if built to scale, but it could also be selectively compressed in a space as small as 4 x 8 feet.

The advantage of this kind of layout is the intensity of operation provided in a small area. I once worked at a paper mill in Wallula, Wash., that at times operated two small 0-4-0 diesel-mechanical switchers, as well as a Union Pacific EMD GP9 that moved strings of wood chip hoppers to and from the chip-unloading staging yard, boxcars to and from the paper-loading warehouse, and tank cars in and out of the boiler and chemical-recovery area. And this didn't count the mainline freight trains that rolled by every hour or two on the main line.

The Oregon, California & Eastern, owned by Weyerhaeuser, supplied the company's mills in Oregon's Klamath Basin. Here, two of the railroad's General Electric U25B diesels haul a long string of log flats in 1977.

Logging locomotives

The earliest recorded photographs of purpose-built logging railroads show trains powered by older mainline locomotives that had been sold or leased to the logging operators. This was a pattern that would persist for most of logging railroad history. Only in logging's declining years was the majority of steam locomotives used in the woods built specifically for that purpose.

Not every old standard-gauge locomotive was suitable for a second life on a logging line. Weight and flexibility were the keys to success in running a locomotive on the wiggly, uneven track found on most logging lines. Late 19th-century loggers used many old 4-4-0s, especially in southern and mid-Atlantic states. In general, older 4-4-0s were flexible and light

Northern Redwood Lumber Co. No. 5 is an excellent example of a later model three-truck, 70-ton Shay with cast trucks and an all-steel cab. *Edw. R. Kirkwood*

Weyerhaeuser's Oregon, California & Eastern GE U25Bs, acquired second-hand, were big locomotives for a logging line, but the railroad's steep grades and big trains demanded lots of power. This view is from 1981.

enough to run on the light (56 pounds per yard) rail typically used by loggers. Small, light 0-6-0s and some very small 0-4-0s also show up in early photos, but many of these appear to have been built as industrial locomotives and may not have been downgraded mainline engines.

Early 1870s catalogs of locomotive builders Porter, Hinckley, and Baldwin show several variations of small, light locomotives "deemed useful for mining, log hauling, and other industrial uses," according to the Porter catalog.

Conventional rod locomotives were the mainstay of the logging industry. Rod locomotives get their name from the use of large connecting (main) rods from the pistons to one set of drivers and the accompanying rods that connect the drivers to each other (side rods). The beautiful synchronized dance of the main and connecting rods and the outside valve gear is akin to the finest ballet for the lover of steam locomotives.

The first commercial Shay geared locomotives appeared in the early 1880s, but didn't see widespread use until the late 1880s. The two other popular

builders of geared locomotives, Climax and Heisler, entered the business soon after Shay and Lima teamed up.

The characteristic common trait of all these early locomotives – rod and geared – is their small size. The few large Shays that appeared prior to 1900 were usually specialty locomotives for mainline carriers.

To understand the different uses of rod and geared locomotives, think of rod locomotives as trucks and geared locomotives as bulldozers. With its low gearing and tracks, a bulldozer can go almost anywhere, but you wouldn't want to drive one from Chicago to Denver!

Rod locomotives can travel at higher speeds, which was critical to a company's profitability as cutting areas moved farther away from the sawmill. However, the trade-off was that compared to geared locomotives, rod locomotives were less flexible, required better track to effectively use their higher speed, and couldn't climb steep grades.

Geared locomotives were much slower, but they were powerful and flexible. Their lower gear ratio (compared with the direct drive of a rod-connected locomo-

tive) meant that more locomotive maintenance per mile of operation was required. Also, a Shay, Heisler, or Climax would burn two to four times as much fuel per mile compared with an equivalent rod locomotive.

For these reasons, rod locomotives were used where possible and practical. Geared locomotives were generally used in the woods at the cutting areas, with rod locomotives assigned to the longer hauls.

Fuel and sparks

Many logging locomotives in the forests burned wood for fuel, especially in the industry's early years. Some locomotives remained wood-burners until the day they retired, with some lasting into the 1950s. However, for the most part, as locomotives grew larger, the fuel changed too.

In the Appalachians, coal – with four times the BTUs compared to a given amount of wood – became the preferred fuel for larger logging locomotives. Coal produced sparks, but fewer sparks than a wood-burner, and their sparks could be controlled reasonably well with screens in the smokebox and with specialized stacks.

In the West and Southwest, oil was the fuel of choice in the woods. Although more expensive than wood, oil produced far fewer

Steam locomotive models

Scale	Description	Mfr.	Comments
N	2-truck modern Shay	Atlas	Excellent. The only prototypical logging locomotive in N scale.
HO	3-truck early Shay	Bachmann	Excellent, especially for 1900 to the end of steam logging era.
HO	2-truck Climax	Bachmann	Excellent. Straight boiler makes it a rare prototype.
HO	2- and 3-truck Heislers	Rivarossi	Excellent. They run well for being designed in the 1970s.
HO/HOn3	2-truck Shay	MDC*	Been around for awhile, but still holds up well.
HO/HOn3	3-truck Shay	MDC*	Been around for awhile, but still holds up well.
HO	2-6-6-2, 2-6-6-2T	Mantua	Generic.
On30	Small 2-truck Shay	Bachmann	Excellent.
On30	2-4-4T Forney	Bachmann	Close to Maine Forneys used for logging.
On30	2-truck Climax	Bachmann	Excellent. Among the most accurate On30 models made.
F (G gauge)	36-ton Shay	Bachmann	Excellent, but wheels not to scale standards.
F	25-ton Climax	Bachmann	Excellent, but wheels not to scale standards.
F	2-Truck Heisler	Bachmann	Excellent, but wheels not to scale standards.

* MDC – Model Die Casting (Roundhouse), now part of Horizon Hobbies

sparks and made the life of a fireman, especially on large locomotives, much easier. Baldwin's big ariticulateds were almost exclusively oil-fired. Only a few burned wood, and then only early in their careers.

Almost all steam locomotives operating in the woods needed some sort of spark arrestor over the exhaust stack, even when burning oil. The stack most commonly associated with wood-burning locomotives was the Radley & Hunter. This stack looks like an inverted cone on the smokebox with a less-angled cap over the top.

Whenever you can see a significant portion of the straight stack pipe, but with some sort of large apparatus at the top, you are looking at a spark arrestor associated with coal or oil. The best example of the most common stack used for coal is the flat-top diamond stack used on Bachmann HO scale Shay models.

Shays and other geared locomotives rarely had true diamond stacks like those found on 19th-century locomotives. The true diamond stack was falling out of use by the time Shays were in production, and Lima offered several other variations of cinder catchers and spark arrestors.

Working the main line for the big Crossett Lumber Co. mill in Crossett, Ark., in 1957, this 2-8-0 hauled boxcars of finished lumber from the mill to an interchange. The locomotive isn't fast, but it can pull – typical of power on a sawmill line. *Matt Coleman collection.*

This Shay has a Radley & Hunter smokestack, which was used on both coal- and wood-burning locomotives. This design was an improvement to the balloon stacks common in the 1860s and 1870s. *Matt Coleman collection*

The availability of the Kemtron (now Precision Scale) HO diamond stack from the Rio Grande C-16 led many unsuspecting modelers to add it to a locomotive to make it a wood-burner. This stack was strictly for coal (the screen area was too small for the amount of ash that wood produced), and it was replaced by better stacks and changes in cinder control soon after the turn of the century. The proportions and sizes of stacks are dictated by steam use and the size of the boiler and firebox.

Modelers often modify existing locomotives with a balloon, diamond, or other stack in an attempt to backdate it or indicate that it burns wood. Look at photographs carefully before you do this, and make sure that you use the correct stack – not only in shape but in size. Most of the common stacks used on logging locomotives are available in HO and O scales.

Second-hand engines

For many logging operations where the cutting and milling might only last one or two seasons, a well-used mainline locomotive would provide better service than animal power. Often these locomotives started out as mainline express locomotives and in their downgrading had elegant wood pilots replaced with a beam and steps, or their large oil headlights replaced with temporarily rigged alternatives, including electric lights from automobiles. Loops of wire rope and chains often adorned these fallen beauties. When detailed with clear reference to the prototype, a model of an aging 4-4-0, carefully and prototypically modified for use in the woods, can be as inspirational as the largest Shay ever built.

Loggers used light rail because it was cheap and could be reused with few problems. Because of the light rail and marginal roadbed construction, more wheels were always better.

After the late 1800s, few used 0-4-0s were seen; their axle loading was just too high. More common were used light 0-6-0s, 2-6-0s, and 4-4-0s. More modern switchers were rarely used, as

Photographers loved the mechanism (or "working") side of Shay locomotives, so images like this one showing the blind (or "off") side of these engines are harder to come by. This is Diamond & CalDor No. 4, a two-truck Shay. *Matt Coleman collection*

This blind-side view of Rayonier No. 10 shows the modern cast-iron trucks of a 94-ton Pacific Coast three-truck Shay. The locomotive was built in 1930 and shown here in operation in 1960. *Philip C. Johnson collection*

by the turn of the century, steam switchers had become heavy, powerful locomotives with high axle loading.

Narrow-gauge loggers had fewer choices when seeking a cheap but good used locomotive. Baldwin and Hinckley 4-4-0s and 2-6-0s are a common sight in old photos, along with Shay and Climax locomotives. Narrow-gauge logging lines, much like their common-carrier brethren, disappeared rapidly in the 1930s, and by the 1950s only a few were left. One of the most interesting was the Argent Lumber Co., of Hardeeville, S.C., which logged coastal yellow pines and hauled trains of their logs with small 2-6-0s and one tiny 2-8-0 into the 1950s.

The turn of the 20th century saw the sudden appearance of street tram locomotives in the woods. In the Pacific Northwest, a number of smaller logging operations were built to 42-inch gauge because of the availability of dozens of 0-4-2T street railway locomotives displaced from the Portland Traction Company as the city lines became electrified.

Two-truck Lima Shay No. 7 of the WM Ritter Lumber Co. in West Virginia has a bent pilot, but is otherwise clean and in good condition. *Matt Coleman collection*

These turned out to be excellent locomotives – small, light, and more flexible than an 0-6-0 thanks to the trailing truck. They lasted many years in the woods hauling logs to mills and landings along the Columbia River.

Grandt Line once offered a conversion kit for its On3 Porter locomotive that could be used with a kit or modified for the Bachmann On30 small Porter. The best source of photos for these are John Labbe's *Railroads in the Woods* and Michael Koch's *Steam and Thunder in the Timber.*

Used locomotives that came to logging lines were quickly modified. The pilot was usually replaced with footboards to provide riding space for the brakeman. Cab windows were enlarged, giving the engineer and fireman better views of the area around the locomotive for safety purposes. (There are more people around log loading and unloading sites than in a yard or industry spur on a mainline road.)

Tenders sometimes received hoses (siphons) for getting water from the nearest stream or pond, and in later years, steam pumps were sometimes added to allow water to be pumped from greater distances.

Some early locomotives had steam winches added to the pilot beam for log loading. This winch, combined with suitably placed blocks and tackle, could load logs onto a log car in areas far from the nearest hoist, yarder, or spar pole. For safety reasons these had pretty much been eliminated by the late 1920s.

Meadow River Lumber Co. No. 7 is a good example of a three-truck standard-gauge coal-burning Shay with a Radley & Hunter stack. *Matt Coleman collection*

This three-truck Heisler was hauling coal for the Moore Keppel & Co. by the time of this 1950s view, but the locomotive had served the same line earlier when lumber was the main industry. *Two photos: Matt Coleman collection*

A pair of Shays rests in a typical backshop area of a logging line. Note the double-drum yarder in the upper left of the photo, and the canvas cab shade on the engineer's side of the Shay at right.

Second-hand locomotives can provide interesting additions to a model logging roster, but they need to be used within the context of the prototype. You're not likely to find a streamlined Hudson hauling logs out of the hardwood forests of Pennsylvania, but you'll find many smaller, older, and lighter locomotives.

Specialty rod locomotives

Soon after the turn of the century, new 2-6-2s and 2-8-2s with low drivers (51-inch diameter and smaller) began to appear in the woods alongside older second-hand engines. These new locomotives were designed with logging in mind and dispensed with some design elements that were the norm for common carrier engines. One feature was the firebox, which on these locomotives was designed more like that of a switcher, with the focus on stability, flexibility, and efficiency instead of raw power.

These new engines were soon in operation across the country and began to change how logging railroads operated, creating a more distinct separation of duties for geared and rod locomotives.

Perhaps the best known locomotives in this class were Baldwin's 2-8-2s with tenders and ultra-low 44-inch drivers. These tiny drivers were possible because of improved design of the coun-

This Caspar, South Fork & Eastern 2-6-6-2, shown in 1948, hauled giant redwoods on a remote California line. *Matt Coleman collection*

terbalances and equalization. Small drivers meant great pulling power, yet they could still run at 30 to 35 miles per hour without damaging the rail, which some rod locomotives tended to do at speed.

Unlike mainline 2-8-2s, the trailing trucks on these logging engines were not designed to support a larger firebox. Instead, they were designed to guide the locomotive when backing, allowing it to run in either direction equally well.

These 2-8-2s, especially in their later variants with piston valves and equal-size trailing and pony (lead) truck wheels, were also quite handsome as locomotives go. It is too bad only a few models have been imported in brass, because they would be popular and useful locomotives for Pacific region modelers.

Alco and Baldwin both built some 2-6-2 (Prairie) locomotives with similar characteristics, one of which operated occasionally on the McCloud River in California until recently. These specialized locomotives were sold from the late teens through the early 1940s to dozens of logging and forest-product railroads. These 2-6-2s were light, could operate on 56-pound rail, ran equally well in either direction, and could

Southwest Lumber Mills kept 2-6-6-2 No. 4, a big Baldwin logging Mallet, in pristine condition. *Matt Coleman collection*

White River Lumber Co. 2-6-6-2T No. 7 hauls a train of logs near Enumclaw, Wash., in 1945. Baldwin built the locomotive in 1925. *Albert Farrow*

Weyerhaeuser No. 9, a 2-6-6-2T, illustrates how the water tank wraps around the side of the boiler. *Baldwin Locomotive Works*

Even though it's a tank locomotive (with the water tank around the boiler), Rayonier 2-6-6-2T No. 111 also sports a tender for additional water capacity. The fitting atop the stack is a Gerlinger spark arrestor. *Whit Towers*

burn oil, coal, or wood. They were Baldwin's answer to the ever-increasing size and power of contemporary geared locomotives – especially the Pacific Coast Shay, the Willamette, and the three-truck piston-valve versions of the Climax and Heisler.

A few 2-4-2s were built for logging, with the double-ended design again being the selling point. Tank locomotives had their supporters too. These distinctive engines had their water tanks wrapped around the boiler instead of in a tender.

The magnificent Alco Minarets, 2-8-2Ts in the Sierras, as well as similar 2-6-2Ts, were used along the Pacific Coast. Many lasted until the end of steam as mill switchers, as their lack of tender gave them flexibility in the confines of a yard.

Another specialty class of locomotives was the 0-4-4T, built in relatively large numbers for the two-foot-gauge Sandy River lines in Maine. Although technically common carriers, these railroads hauled more timber and pulpwood than any other commodity. The big Eustis Nos. 20, 21, and 22 were outside-framed, long-wheelbase locomotives that, along with their Phillips & Rangeley sister, No. 17, were designed to haul heavy loads of timber on tiny tracks in the broad valleys of eastern Maine.

The two-foot lines had broad curves and relatively gentle grades, and the Forney style of locomotive answered the need for a large locomotive without the problems of a tender or a tank over the drivers, which can cause

This Berlin Mills 2-4-2T is typical of the smaller steam locomotives used for switching around sawmills and paper mills. *Matt Coleman collection*

Simpson Lumber Co. No. 12, a 2-8-2T with tender, waits at the loading spar. By this 1949 view, the Washington line was a truck-to-rail reload operation. *Fred Matthews, Jr.*

Argent Lumber Co. No. 7, a three-foot-gauge 2-8-0, burns what it hauls: yellow pine from Georgia forests. The boiler backhead keeps coffee and lunch warm. *Jim Shaughnessy*

a dramatic drop in adhesion as water is used during operation.

Articulateds and Mallets

Of all the locomotives that ran in the Pacific Northwest woods, the articulated tank locomotives and slightly larger tender-equipped locomotives were, and are, the most impressive and amazing.

Articulated locomotives have two engines (two sets of drivers below the boiler). The Mallet, developed by Anatole Mallet, was a compound articulated. This meant the steam was used twice: first in the cylinders of the rear engine, then in the forward cylinders. Although some refer to all articulateds as Mallets, only compound locomotives can accurately be called Mallets.

The articulated logging Mallets of the West were developed as the Baldwin and Alco Prairies (2-6-2s) and Mikados (2-8-2s) were being stretched to their limits on larger lines. Most track was too curvy for a 2-10-2, and articulated locomotives had already been proven in other uses by the time Baldwin began to sell them out West. Several 2-4-4-2s were built for logging, but the rest were 2-6-6-2s and 2-8-8-2s. The 2-6-6-2s

were the most common and came in tank (saddle and side tank) versions and as tender locomotives. Models have been offered in brass in HO and O scales, and Mantua offers a die-cast version that, while not completely accurate, can be detailed to match several prototypes.

The big railroads, including Rayonier, Weyerhaeuser, Long-Bell, Bloedel-Donovan, Hammond, and Kosmos Timber, hosted the giant articulateds that arrived on the logging scene late in the game. Built during a short

period from the 1920s through the 1930s, these locomotives – some built as simple articulateds, others as true Mallets – put many smaller and older locomotives out of service.

Geared locomotives

Geared steam locomotives are known as indirect drive locomotives because the main rod, coming from the steam cylinder, is not connected directly to the

Wood-burners need three to four times as much fuel as coal burners. Here, Argent No. 2, a three-foot-gauge 2-6-0, has a full load in its tender. *Jim Shaughnessy*

Maine two-foot-gauge Sandy River & Rangeley Lakes was a thinly disguised logging railroad. When the timber ran out, the railroad did the same. Here No. 9, a Forney locomotive, hauls a mixed train with a pulpwood load leading. *Matt Coleman collection*

crank axle on the drivers. In the case of the Shay, the engine (with two or three cylinders) is on the side of the boiler and powers a longitudinal drive shaft. The shaft turns bevel gears, which in turn move against larger ring gears on the drivers. The Willamette was simply an improved Shay that was constructed after the original Shay patents ran out.

The Climax and Heisler, with their cylinders located in the center, drive a central shaft (very similar to the Shay), but the shaft is centered under the boiler and connected to the drivers through gearboxes on the axles. On the Heisler, the gearbox is only on one axle per truck, with the other axle driven by a standard steam locomotive connecting rod.

There were also chain-drive locomotives (the Sykes), and Baldwin built a geared locomotive that had many characteristics of the Climax. There were other makers of geared locomotives, but none could match the existing builders' capacity, and in some cases, they were simply so close to the others that there was no practical difference. The only minor builder to challenge the big three was Willamette, a company already well established as a builder of steam

yarders, winches, and other logging equipment.

Photographs of geared locomotives abound. One excellent Web site is Geared Steam Locomotive Works (www.gearedsteam.com), and there are sites for individual builders as well. Books have been published covering all of the major geared locomotives. *The Shay Locomotive – Titan of the Timber* by Michael Koch is the definitive work on the Shay (albeit long out of print). The Climax, Heisler, Willamette, and Dunkirk have all had individual books published on them as well.

The Shay

The Shay Patent Locomotive was the first geared locomotive designed for logging, and it was also the most successful. The Shay was the product of the mechanical genius of Ephraim Shay of Michigan. Shay, a logger, noticed that a flatcar heavily laden with logs could traverse track that would cause even the lightest of his locomotives to derail. The locomotive's large drivers (compared to the log car) and the pounding motion of the side rods were contributing factors.

In the early 1870s, Shay began looking at ways to solve the problem. He eventually settled on

The Eustis Railroad, a Maine two-footer, operated three outside-frame Forney locomotives including No. 8. *Matt Coleman collection*

a rather conventional boiler with a small vertical steam engine mounted alongside the boiler, with the engine driving a crankshaft through universals to a bevel-and-ring gear arrangement on each wheel.

This design, although seemingly complex, survived during the entire Shay production run. The design was loved by logging railroad mechanics, who could repair almost any moving part on the locomotive without having to crawl under the engine or use a crane. This maintainability, coupled with the rugged reliability built into Victorian-era railroad equipment by all builders, made the Shay incredibly enduring.

Shays came in many sizes and had two primary markets, logging and mining. Some of the largest Shays worked on coal branches in the Appalachians, but some ended up on logging lines.

In 1907, cast engine-bearing frames came into common use and the T-boiler era ended. (This was a straight-course main boiler section that fastened directly to the cylindrical firebox, with the top of the firebox taller than the boiler and the ashpan below the

A diminutive Porter 0-4-0T works in the great north woods of Minnesota. Although not powerful, small Porters like this one enabled small logging operations to get in business quickly and cheaply. *Library of Congress*

boiler section). The Web site www.shaylocomotives.com provides lots of information and details.

While most Shay models seem to focus on the larger and later versions, most Shays weighed 42 tons or less, and most were the two-truck version. Some of these were still in operation well after the last of the big Pacific Coast Shay locomotives were retired.

Heisler

The Heisler Locomotive Company manufactured its unique design of geared locomotives from 1892 until 1941. The Heisler was well received by the logging industry, with 850 locomotives built. Heisler Locomotive Works opened as the Stearns Manufacturing Company in Erie, Pa., changing its name in 1907 to match the most important of the company's products.

The Heisler locomotive had several unique features. A V-twin steam engine powered a central drive shaft that in turn drove a single gearbox on the inside axle of each truck. Rods connected the powered drivers to the other pair of drivers on each truck. Heisler claimed that its balanced drive was smoother than the Climax, and the truck equalization and drive shaft connections made the Heisler the most flexible geared locomotive. It was also faster (but only marginally) than the Shay, and debate still rages as to whether the Climax (18 mph) or the Heisler (16 mph) was faster in real service. Despite their somewhat awkward appearance, Heislers were widely accepted

Sumpter Valley No. 18, built by Baldwin, was a modern-design, low-drivered, inside-frame 2-8-2 narrow-gauge wood-burner. *Matt Coleman collection*

Many Civil War-era 4-4-0s finished their careers far from the varnish of premier passenger trains, working logging lines in flatlands or valley bottoms or as mill switchers. *Library of Congress*

in both the logging and mining industries and served to the end of steam.

The success of the Heisler, especially the large three-truck versions in the Pacific Northwest, made it a mainstay of the final logging operations in Washington. The Geared Steam site (www. gearedsteam.com) is probably the best Web site for Heisler photographs and information.

Climax

The Climax, built by the Climax Manufacturing Co., of Corry, Pa., is one of the most modeled of all the geared locomotives. Its design originated by combining a small two-cylinder marine steam engine with geared trucks. This original design, with an enclosed cab and the engine and mechanical works hidden from sight, was known as a Class A Climax. Some had vertical boilers, but the T-boiler, as in the Shay, was the primary early boiler. A few were built as 0-4-0s but most were built as double-truck versions.

The design soon changed to the more familiar version with a pair of cylinders inclined on both sides of a horizontal boiler. The cylinders, through a transfer gearbox, turn a central shaft that goes to the double gearboxes on the trucks. This design allowed additional gearing options. Some Climax locomotives were built with two-speed gearboxes, allowing a higher transit to help overcome the limitations of running geared locomotives for long distances.

Excellent scale versions of this locomotive type are available in brass in several scales. Many of the HO and O scale conversion kits are excellent models in their own right and if modeled with a true Climax chassis would result in excellent scale models. David Hoffman of The Bald One Locomotive Works makes a complete Class A Climax kit in HO narrow gauge that is quite accurate.

In addition to the books on this locomotive, there's a Web site

devoted to Climax locomotives (www.climaxlocomotives.com).

Willamette

The history of the Willamette locomotive is complex but fascinating. It is told in excellent detail in *The Willamette Locomotive* by Steve Hauff and Jim Gertz.

Willamette Iron and Steel Works, of Portland, Ore., had been active in the Northwest since its founding in 1865. By the turn of the century, it was a major builder and repairer of logging equipment of all kinds. The company often did repair work on geared locomotives that had been damaged in wrecks, and its ability to fabricate almost any part of a locomotive made the company invaluable to loggers. Willamette also manufactured large steam-powered yarders, skidders, towers, and other components of the modern high-lead logging systems used in the Northwest (shown in chapter 5), and the company had the capacity to manufacture almost any component loggers needed.

As the industry matured, it was clear that big-time loggers were looking for a more modern locomotive that would be less expensive to maintain yet more economical to run. Many companies had gone to oil-firing of their locomotives to reduce the fire hazard during the dry seasons and to reduce the labor costs associated with wood as a fuel.

Willamette developed a prototype geared-locomotive design in 1921 and began testing the first two-truck Willamette in November 1922. Designed with a superheater, the locomotive had many modern features and the efficiency and safety devices of contemporary rod locomotives.

Over the next eight years, Willamette built 33 locomotives, each one with improvements and enhancements. Piston valves followed soon after the first test locomotive, and three-truck versions emerged. The locomotive was well accepted by loggers of the Northwest.

Although its production life was cut short by the onset of the Depression, Willamette's venture in locomotive manufacturing threw down the gauntlet to the

This little 0-4-2T worked in building the Alaska Railroad, then went on to serve two logging operations along the line. *Library of Congress*

Two General Electric 70-ton diesels lug 70 loaded log cars along Weyerhaeuser's ex-Southern Pacific branch out of Springfield, Ore., in September 1979. *Carol L. Ingles*

other three builders. All responded quickly with versions of their own locomotives that essentially copied what Willamette had done. Although strictly a West Coast locomotive, the Willamette is an important modern logging loco-

motive. Brass models of various Willamette engines have been imported in O and HO scales.

Other geared locomotives

Numerous other limited-production geared steam locomotives

Five assorted Electro-Motive switchers lead 65 log loads to Weyerhaeuser's mill at Longview, Wash., in 1986. *John C. Illman*

33

Taking pride in his charge, the engineer takes advantage of some down time and waxes the finish of Weyerhaeuser GP9 No. 765 as it waits to unload logs at the South Bay Dump at Henderson Inlet, Wash., in 1980. *Gordon Lloyd, Jr.*

The boiler on each Shay was offset to the side to compensate for the weight of the engine on the locomotive's right side.

were designed for logging. Some were reasonably successful; others were not.

Seeing the size of the geared-locomotive market, Baldwin built its own geared engine. Although limited in success (only five were built), it helped Baldwin learn what the logging industry needed and indirectly improved the designs of its rod locomotives.

Some railroads built their own locomotives. Adaptations of boilers to railroad chassis were within the capability of most skilled mechanics of the logging era, and some of their creations are wonders to behold.

Many of these homebuilt locomotives were chain driven, with the Sykes (with four or five built) the best known among those who built more than one. Some of these lasted a decade or two, which indicated a soundness of design, if not efficiency.

Diesels and electrics

Logs and wood products continued coming out of the woods by rail long after the last boiler went cold in the woods, hauled by diesels of the same makes and types used on common carriers.

The Red River Lumber Co. used a oil-electric built by American Locomotive Co. (Alco) and GE as early as 1926, and it was a contemporary of the giant logging Mallets and three-truck geared locomotives from the heyday of Pacific logging.

More common were smaller gas-mechanical switchers used around sawmills to move cars of lumber to drying stacks and to the shipping dock. These didn't need a fireman and were more economical to operate and maintain than small steam engines. By the 1940s, they were common in most regions of the country.

Gas-mechanical speeders and even some old McKeen railcars found their way into the woods to move loggers from their bunkhouses to the woods more quickly and efficiently. Automobiles and small trucks were converted to run on rails, and they began

Former West Side Lumber Shay No. 7, a three-foot-gauge locomotive, rests on display in this 1981 view. It had been in service on a tourist line.

Willamette geared locomotives, built for service in the Pacific Northwest, were large engines. They were similar in design to Shays, but Willamette locomotives had their piston valves all facing forward, a feature meant to be an improvement over earlier Shays.

to replace the odd boxcar or enclosed delivery car that used to accompany some logging trains. These converted cars also had the duty of carrying a doctor to the site of an accident or, more likely, bringing an injured logger to the camp hospital.

The Red River Lumber Co. not only pioneered the use of big diesel-electrics for logging, it even experimented with using electric locomotives to haul timber out of the woods. Using a unique side-trolley system to keep overhead clearances free for moving oversize loads, the 1,500-volt system was successfully used for many years. A few photos show log cars being hauled behind some Portland Traction freight motors on the line southwest of Portland.

In West Virginia, the Meadow River Lumber Co. used a General Electric 70-tonner for a few years hauling logs on skeleton log bunks from the reload site down to the mill, providing an excel-

lent prototype for a transition-era layout.

Diesels took over on the remaining reload or mill-to-interchange lines after steam locomotives were retired. Coos Bay Lumber EMD SW1200s with dynamic brakes, GE U25Bs on the Weyerhaeuser line near Klamath Falls, and the Baldwins on the Amador Central are all good examples.

Geared steam models

Whether prototype or model, geared steam locomotives, particularly Shays, are amazing sights in operation. With the Shay, and to a lesser extent the Climax and Heisler, movement is compounded and in slow motion. On the Shay, the three cylinders each have their main rods connected to the crank and each has valve gear in motion. The rotation of the crankshaft, together with the individual articulation of each truck, is an absorbing sight. The slow speed, the noise, the motion,

and the sense of controlled power make the Shay among the most appealing of all locomotives.

Geared locomotives have been offered in some form in every scale from N through large scale. Models of logging locomotives followed the growth of the model railroading hobby, and by the mid- to late 1950s Lobaugh offered an O scale kit for a two-truck Climax, and Max Gray had imported a brass O scale Shay.

Although crude by today's standards, these models touched on a segment of the O scale hobby that had never been served before. Pacific Fast Mail (PFM) began to import hand-made brass Shays in HO scale, and Kemtron introduced its famous line of castings for a Shay in O narrow gauge.

In the 1970s, AHM/Rivarossi brought out plastic models of two- and three-truck Heislers. These became mainstays of many a logging layout because of their reasonable cost and excellent running capabilities. Then Bachmann

This cosmetically restored Class C (three-truck) Climax locomotive was on display in Lewiston, Idaho, in 1982.

introduced its watershed 36-ton Shay in large scale (1:20.3) in the 1990s, followed by a three-truck Shay in HO. Many other models followed, and we now have inexpensive, smooth-running Shay, Climax, and Heisler models in many scales and gauges, including the beautiful Atlas two-truck Shay in N scale.

Older brass locomotives are available through dealers and on eBay. These brass models are beautiful and well detailed, but be aware that their running qualities aren't as good as a modern HO locomotive model. Many have older open-frame motors that don't run as smoothly and aren't as easy to convert to to DCC as can-type motors. Also, the running condition of these locomotives may not be known until the locomotive arrives at your door.

Modeling guidelines

As tempting as it might be to take any small steam locomotive, cover it with detail parts, weather it, and call it a logging locomotive, such a model likely won't be accurate.

Also keep in mind that although there are few absolutes in logging, especially when it comes to locomotives, modeling the common instead of the unique will result in a more realistic layout. Many exceptions and unique situations are well documented. But these exceptions are just that – exceptions. I encourage people to avoid modeling exceptions unless they are trying to capture the specific prototype than ran them.

Today we have more accurate, smooth-running logging locomotives than ever before. We also have more books, videos, Internet sites, and other sources for accurate prototype information. By using the information resources available, combined with the high-quality equipment that is available, it's easier than ever to capture the wonderful mystique that is a logging railroad.

General Electric's 44-tonner is a diesel locomotive well-suited for service on light branch lines, a job No. 4 served for Weyerhaeuser in Oregon.

Rolling stock and track

ogging lines operated a wide variety of rolling stock, much
of which isn't familiar to common-carrier railroads. This
chapter takes a look at this equipment, along with track that's
appropriate for a logging railroad.

Unique equipment

Rolling stock on individual logging lines varied greatly – some
railroads operated specialized cars that were truly one-of-a-
kind, built for a specific purpose and used for nothing else. But
in general, logging rolling stock had many similarities in
construction and use when compared with their common-
carrier brethren. In later years, as cars became more specialized and built to higher standards, they were often sold among
logging railroads. Thus, equipment from one railroad often

The Louisiana Cypress Lumber
Co. used these skeleton cars
until the end of operations in
the 1950s. The cars had
archbar trucks and link-and-pin
couplers. Note the uneven
track, with no tie plates
between the rails and ties.
C.W. Witbeck

FOUR

Speeders such as this Oregon, California & Eastern car were common after World War II. They were used for maintenance as well as for following log trains for fire protection.

Weyerhaeuser kept this brightly painted modern steel cupola caboose in service into the 1980s on its Oregon, California & Eastern line. One of the line's U25B diesels is at left.

ran on other logging lines during its life.

In the world of common-carrier railroads, there are specific distinctions for equipment based on whether or not it contributes directly to earning money. Rolling stock is called "revenue" or "non-revenue" depending on its use and its accounting classification.

Maintenance equipment, including cars for work trains, cranes, pile drivers, ditchers, flangers, snowplows, and cabooses, are typically classified as non-revenue cars. But on a pure logging line, which exists only to haul timber from the forest to the sawmill, there is no such distinction.

By the 20th century, many pieces of logging equipment – log bunks or racks in particular – were not authorized for interchange because of the different locations (or absence) of items such as grab irons, brake wheels, and retainer valves (on equipment with air brakes). As a result, logging railroads tended to be unique, self-contained operations. Because most logging lines were not common carriers, they did not interchange cars with other railroads. Cars were pulled to the cutting area, loaded, brought down to the sawmill, unloaded, and then sent back. The finished, cut lumber was shipped out in common-carrier boxcars that

never ventured onto the logging tracks. Often, even when the sawmill shipped finished products out over the logging company lines to an interchange, different locomotives were used for each operation as noted in chapter 3.

There were instances of common-carrier equipment on logging lines and vice-versa, but these cases almost always occurred before 1908 (the year comprehensive railroad safety laws took effect nationwide) or after 1970, when logging railroad equipment came under the same railroad safety laws and many steel-framed logging cars were built or rebuilt to interchange standards.

Also, some equipment used on logging lines, such as fire trains, water cars, and speeders, were sometimes mandated by state or federal agencies during fire season or for safety reasons.

Weathering and aging

Don't fall into the trap of weathering and distressing models of logging equipment until they look as though they would barely run. Locomotives and cars on logging railroads were the bread and butter of the operation. They were maintained in good condition even if structures in the surrounding camp were not. Although equipment in the woods rarely looked like it had just come out of the paint shop, it still should appear well maintained and serviceable. Some companies had a standing rule for all locomotives to be washed when they went through the enginehouse for routine repairs, so a clean, shiny locomotive would not necessarily be out of place.

Log cars were also well maintained, with regular shop visits, especially on larger operations. For seasonal railroads, such as the West Side Lumber Co. in later years, the off-season was when they repaired, maintained, and painted all the equipment.

Models of logging rolling stock

Scale	Description	Mfg	Era	Comments
HO	70-foot wood chip gondola	Walthers	1980-	Accurate
HO	70-foot wood chip gondola	LBF	1980-	Accurate
HOn3	Log car	MT	1910-1950	Freelanced design
HO	Log car	Kadee	1910-1950	Accurate/excellent
HO	Truss-rod log car	Kadee	1900-1950	Accurate/excellent
HO	Disconnect trucks	Kadee	1895-1940	Accurate/excellent
HO	Caboose on disconnect truck	Kadee	1895-1940	Based on prototype
HO	40-foot log flat	Bachmann	1900-1930s	Freelanced with prototype features
HO	Skeleton log bunks	Rivarossi	1900-1950	Accurate/excellent
HOn3	New Mexico Lumber Co.	RGM	1890-1940	Accurate/excellent
On30	Skeleton log bunk	Bachmann	1890-1930	Freelanced; shortened from 3-foot gauge prototypes
HO	Skeleton log bunk	Keystone	1890-1930	Short car, based on Eastern prototype
HOn3	West Side 24-foot flatcar	RGM	1885-1940	Accurate/excellent
HO	Climax flat	Keystone	1890-1940	Accurate/excellent
O	Wood chip car	MTH	1970-	Accurate, but needs scale trucks
N	Wood chip car	DeLuxe	1970-	Accurate
HO	Grasse River caboose	Keystone	1900-1940	Accurate/excellent
HO	Barnhardt Log Loader	Keystone	1900-1950	Accurate/excellent
HOn3	West Side short caboose	RGM	1910-1961	Accurate/excellent

Manufacturer key: MT – Micro-Trains; RGM – Rio Grande Models

Buildings were another matter. Some were simply unpainted, with the wood taking on a gray, weathered hue. Most structures, especially buildings in the forest camps, were well maintained and painted.

To judge for yourself, look to the prototype first. Try to track down good in-service photos of the equipment you are modeling.

Log cars

The earliest log cars were flatcars from the local common-carrier railroad. However, whether borrowed or leased, mainline railroads quickly realized that their equipment, once taken into the woods, never came back quite the same. Documents from the Savannah & Atlanta Railroad in the 1860s list regulations against letting any of its equipment be allowed off the line and onto the lines of any "railroad engaged in timber cutting or drayage."

Logging and loggers were rough on equipment. In the days

Logging cabooses ranged from large to small and primitive to modern. This four-wheel wood caboose served Weyerhaeuser early in the 20th century. *Harold Borovec*

of lightweight truss-rod wood cars, a flatcar loaded with 10 or 12 moderately sized logs could easily exceed the capacity of the car and its trucks. The wood deck and side pockets were often damaged from shifting logs during transit.

As the logging industry grew, companies began to purchase and even build their own cars. Since most logging operations were directly connected to a sawmill, it was easy enough to have the necessary wood cut as needed and

Rayonier's home-built "outhouse-on-wheels" cabooses were signature pieces of equipment. *Stan Kistler, Philip C. Johnson*

buy the iron components from outside suppliers. Car builders eventually began to offer log bunks and flatcars designed for logging railroads.

Disconnect trucks

Disconnect trucks became popular for hauling logs in the 1880s, and some remained in service into the 1950s. Each disconnect consists of an archbar truck with couplers on both ends. The truck is topped by a swiveling bolster and integral log bunk. Each end of a log load rides on a disconnect.

The advantage of disconnect trucks was that they could be used with any length of log, which was especially important when hauling logs that were not cut (bucked) to length. From the 1880s through the 1930s, some logs were kept full length for use as pilings and dolphins in docks and harbors as well as for ship masts, specialized trestle bents, and some architectural uses. Some timber in the Pacific Northwest is still kept at full tree length for specialized purposes.

Disconnects were a compromise in many ways, and although common, they were not well loved by logging companies or loggers.

For one, disconnects were light and would tip over easily on curves, especially when hauling long, unconstrained loads. They were a nightmare on long downgrades because they had only hand brakes, which were located alongside the truck sideframe.

Memoirs of loggers who worked with them describe many hazards of disconnect duty. One was the problem of trying to move among them to set brakes. Another was the danger of leaning over the side of the logs to move the brake wheel to adjust brake pressure. Some logging railroads simply ran trains of disconnects so slowly that brakemen could walk alongside the cars to adjust the brakes, but even that had its dangers if a log load broke loose.

The final problem was that you couldn't push a trainload of loaded disconnects: They would just buckle or twist on the track and then derail. As a result, the trackage at most sawmills that used disconnects was set so that trains were always pulled to the unloading area. A steam winch (usually an older, smaller single-drum yarder) pulled the empty disconnects to a track where they would be connected to each other

and pulled gently back up to the loading area in the woods by a logging locomotive.

Disconnects could be connected by a single beam with a coupling on each end (commonly known as a rooster) for more stability. This was a holdover from link-and-pin days, and the couplers on most disconnects had a slot where a link could still fit, with a hole cast into the coupler face for a pin.

Modelers love disconnects, and good models are available from N through large scales. Keep in mind that although most models can be pushed without much of a problem, anyone familiar with their real-life use in the woods would wince at the sight.

Skeleton cars

The log bunk (also known as a skeleton car) became popular in the early 1900s. Log bunks can be thought of as flatcars without the deck, with heavy center beams (and sometimes side beams) between the trucks. Lines that hauled heavy, large timber ran skeleton cars with truss rods for added capacity.

Kadee makes both types in HO, and both Bachmann and Riva-

Skeleton cars, such as these on the MacMillian-Bloedel line in British Columbia, became the predominant style of log car style by the 1940s. *Jerrold F. Hilton*

rossi offer excellent log cars. Micro-Trains offers skeleton cars in N scale. Craftsman kits of both types have been made in limited runs in almost every scale.

Log bunks were a major step forward in technology because they allowed the use of air brakes. This dramatically improved safety and speed of operation, giving the engineer greater control over the train. Operating at faster speeds increased the railroad's overall capacity. The larger California narrow-gauge logging lines were all air-brake users relatively early, and post-WWII safety regulations forced almost all logging operations to convert to air brakes wherever feasible (disconnects were the obvious exception).

Steel-framed log bunks began replacing wood versions in the 1930s. The Biles-Coleman line in eastern Washington had some of the first steel log cars made for narrow gauge. These were the precursors of today's specialized log cars that are fully interchangeable on common carriers.

The eras of the earlier equipment do blur together. Flatcars, disconnect trucks, and skeleton log bunks can be seen together in photographs on the Pacific loggers, as well as on Sierra pine loggers and West Virginia lines until the early 1930s.

Steel-framed flatcars appear on some Southern lines after 1923, especially those that brought pulpwood, plywood peeler logs, and saw timber from the pine forests of the central South. Some of these were shipped over common carriers, requiring steel flatcars with cast trucks (some with bulkheads) for interchange.

However, you won't see modern steel log bunks of the 1980s together with truss-rod wood cars. Once the log-hauling railroads came under the same laws as common carrier railroads during the late 1960s and early 1970s, the standards for equipment became the same: steel underframes, steel ends, air brakes, positive load control, and standard safety appliance positioning.

Some logging lines made log cars by cutting down old 40-foot boxcars to just the underframe and bracing, with steel side posts, in the 1960s and '70s. These cars,

Weyerhaeuser's log cars were typical of heavy duty cars of the late steam and early diesel eras. They rode on Andrews cast trucks, were equipped with modern air brakes and knuckle couplers, and had drop-down brake wheels. *J. David Ingles*

Specialty flatcars were often put into log service, as here on the Northern Redwood Lumber Co. near Korbel, Calif. These have archbar trucks and truss rods. Note the drop-down brake wheel on the car end. *Two photos: Edw. R. Kirkwood*

when modeled from photos, make distinct and interesting cars.

Modern log cars are most commonly owned by forest product companies or specialty shipping companies serving the industry. In the latter part of the 20th century, the value of larger logs, especially high-quality (furniture-grade) hardwoods, has reached the point that the typical shipping distance from cut to user increased from about 50 miles in 1945 to over 400 miles in 1995.

Modern log bunks are all steel, have roller-bearing trucks, and the side posts (either fixed or hinged) are solid steel and quite massive. They have steel ends to prevent the load from moving off the car in case of a sudden stop or derailment. These cars operate quite widely in the South and Midwest, although rarely more than one or two to a train. They can be considered a specialized version of the modernized reload-line logging cars from the 1970s.

Pulpwood and wood chip cars

The paper industry is a large but less visible part of the forest products industry. In the 1990s,

Walthers brought out a series of HO kits that represent paper mill structures. These kits are very good for representing paper mills from the 1940s through the present, and could easily be backdated into the 1920s and '30s.

The first cars modified and developed to haul wood to paper mills were known as pulpwood flats or pulpwood racks. In the days of wooden cars, these were flatcars with open racks that looked almost like a stock car without a roof (see the photo atop page 30 for an example). Pulpwood (typically six to eight inches in diameter and four or five feet long) was hand-stacked in the cars. These were common on Maine's two-foot gauge railroads, and most standard-gauge lines in Maine and eastern Canada had similar (but larger) cars.

Pulpwood is most commonly used for making newsprint. These plants are common to the northeastern United States and eastern Canada, as well as in the Southeast and upper Midwest. After World War II pulpwood was usually carried on flatcars with bulkhead ends and floors that

sloped inward to the center. This stabilized the load without requiring side constraints, but also restricted cars to speeds under 40 miles per hour in most areas. Pulpwood today typically has a maximum economic shipping distance of about 100 miles.

Some modelers use common 40-, 50-, or 60-foot steel bulkhead flats for pulpwood, but knowledgeable modelers will quickly spot the error. Pulpwood flats don't have stake pockets, and the steel floor is sloped to the center with drain slots in the center. The cars carry two rows of pulpwood, which is usually five feet long, and the slight angle toward the center is clearly visible. Some commercial model pulpwood loads have this detail and can make an ordinary bulkhead flat resemble a pulpwood car.

Specialized wood chip cars also travel between sawmills and paper mills. Modern wood chip cars are 60 feet long and have one end that is hinged, so they can be unloaded by tipping them up on a hydraulic lift.

Before these large wood chip hoppers became the standard in

The steam skidder, with the slack lines, has yarded the logs onto the deck in the foreground. From there they have been rolled onto the log car at right. This 1910 scene is at the Thomas Lake Camp in Washington. *Library of Congress*

the 1960s, chips were hauled in hopper cars modified with extended sides to increase the capacity. Wood chips are less dense than coal, so a car can handle more volume without exceeding its weight limit.

The Southern Pacific and some other Western roads hauled chips in 40-foot drop-bottom gondolas with plywood side extensions. These were in use in Oregon into the early 1980s. They were typically spotted at smaller sawmills that chipped branches, off-cuts, and unusable timber, then blew the chips directly into the cars. When a car was full, the mill shipping clerk would call the railroad, and the next local freight would move the car.

This would be an interesting operation to add to a layout – the sawmill wouldn't even need to receive logs by rail. Many mills in the Pacific Northwest during the 1970s and '80s received all of their logs by truck but shipped cut timber and chips by rail.

Specialty equipment

Many pieces of specialty equipment simply cry out to be mod-eled. Keep in mind that a single prototypical model is worth several "sorta like" models when it comes to capturing the atmosphere of a logging line.

Real logging railroads operated a few blacksmith cars like the HO kits TruScale offered back in the 1950s, but they were not common. Likewise, few prototype cars were built on single trucks. A few cabooses were built on disconnects – Kadee makes an excellent

one – but they were less common than standard cabooses with double trucks.

Logging maintenance equipment was more primitive than that used on other railroads, with most line work done by crews hired in for the job or by loggers during non-logging season.

Some railroads had structures built atop logging flats for transporting repair parts, lubricants, wire rope, and other supplies from the main shops to cutting and loading areas in the woods. (The West Side Lumber Co. had a couple that have been preserved.) These did not run every day, or even every week.

Firefighting equipment

They weren't the most glamorous of the specialty cars, but water cars served two purposes. First and foremost, they were essential for putting out minor fires, such as those from locomotive sparks, before they became major fires. The water car made it possible for what is essentially a portable fire (the steam locomotive firebox) to operate in forests that are highly flammable at certain times of the year. When steam locomotives were the only practical way to get logs out of the woods, fire safety was critical.

Disconnect trucks were popular into the early 1900s, as they were inexpensive and could easily be built by the logging company. *Darius Kinsey, Library of Congress*

Workers load piling onto flatcars for a trip on a common-carrier railroad in 1899. Long loads like this span two cars. *Darius Kinsey, Library of Congress*

A Rayonier Shay eases a 100-foot-long spar tree, spanning two log cars, up a spur. The log will be used for a new loader. *Stan Kistler*

These side-dump cars on the Louisiana Cypress Lumber Co. help keep the railroad's line through the swamp stable by dumping bark and wood chips as fill and ballast. *C.W. Witbeck*

Water cars also hauled water for steam yarders and loaders, especially in drier sections of the forest or when the yarder or loader was on a ridge and too far from a local water supply for pumps or siphons to work.

Water cars were a practical necessity on most of the larger lines – a fire could destroy the entire operation and take lives. During the dry season, water cars were often hauled behind the locomotive tender. In later years, specially built cars had steam- or electric-powered pumps that allowed an operator to direct a stream of water a hundred feet or more from the roadbed to put out minor blazes. When forest fires broke out, logging railroads sometimes ran fire trains to help contain a fire or to protect critical bridges or trestles.

Water cars on early operations were simply barrels of water carried on a flat car. Later, wooden and finally iron and steel tanks appeared. The West Side Lumber Co. had its famous "Coffin Tank" on one of the original Carter Brothers 24-foot flat cars. Rayonier and Weyerhaeuser had specially built tank cars with pumps and operator stations. These lines sometimes ran speeders behind trains with a small water car to watch for sparks from brakes or the locomotive.

Most Western states had fire-control laws by the 1920s or '30s. These mandated water car operation when the fire danger was assessed to be high enough. The U.S. Forest Service had separate regulations for logging operations in national forests, and the Canadian provinces all had fire train regulations. The Forest Service worked with logging companies regarding firefighting, and logging company employees often fought forest fires when they occurred. The railroads' tanks provided a source of water for the firefighters when needed.

Rayonier No. 111 pulls a string of skeleton cars into position at Crane Creek Transfer as log semis await unloading. *Stan Kistler*

Cranes and pile drivers

By the turn of the 20th century, the steam pile driver had became an essential tool for many logging railroads throughout the country. It was almost a necessity for building trestles, which were used in place of fills on the quickly laid logging lines.

In the Pacific Northwest and Michigan, many miles of track were laid across marshy ground using low trestles instead of fill. Some logging companies claimed that they could build track on low trestles using a pile driver as fast as, and for less money than, building up a grade with earthen fills. Where logs were plentiful and waste was not an issue, this was probably true.

As costs rose and times changed, the pile driver was primarily used for bridge repair and trestle construction for line relocation. Steam shovels and early earth-moving equipment made it easier and cheaper to create fills for track by the 1930s.

Steam (and later a few diesel) cranes were also common on smaller operations. Too small for log loading, they were used for picking up after wrecks, which were quite common in logging, and for all kinds of repair and

maintenance duties along the line and in the shop. Photos of small cranes being used to hold Shay crankshafts in place during repair indicate the extent of their usefulness to logging companies.

Railcar-mounted loaders

Car-mounted loaders were similar to cranes. These steam loaders usually featured a two-drum arrangement with a boom, enabling them to load log cars from a siding or main line when there were only one or two cars around the loader.

Car-mounted loaders were usually converted from standard two-drum donkeys placed on specially built flatcars. By using one drum to either move cars on a track parallel to the loader, or to move the loader itself (by setting a line around a solid tree), a car-mounted loader could load a train without needing a locomotive and full crew.

They were used occasionally in the Northeast for hardwood logging and in the Southeast for hardwood and sometimes pulpwood loading. Car-mounted loaders were top heavy, requiring slow speeds when they were moved in trains. Some survived into the 1930s.

The Barnhart loader moved along rails on the log cars as they loaded logs from beside the track. Although more associated with logging in the East, Barnhart loaders were used in almost all regions because of their flexibility. Like a two-drum loader on a flatcar, the Barnhardt could, with proper use of wire rope, move cars without a locomotive. Keystone makes a beautiful model of this loader in white metal. The McGiffert loader was similar to the Barnhardt.

Cabooses

Photographs of early logging operations rarely show any equipment on a train except for the locomotive and log cars with crewmen (presumably brakemen) on top of the loaded cars. Over time, photographs begin to show other types of cars in log trains, especially in winter or on longer, larger operations.

By the 1920s, the larger Pacific Northwest operations were regularly using cabooses; narrow-gauge loggers in the Sierras also had cabooses about that time. This occurred just as air brakes became common on logging railroads and the first of the logging safety laws went into

This turn-of-the-century view shows four-wheeled cut-lumber cars, used to move sawed lumber from the mill to the drying yard. *Library of Congress*

effect in most states. However, photographs of West Virginia loggers – both standard and narrow gauge – don't show cabooses on every train, even as late as the 1940s.

Passenger equipment

Photographs from the first two decades of the 20th century show large logging crews heading into the woods riding on log bunks and flat cars or, occasionally, on old passenger cars. In later years, more gas-mechanical passenger vehicles appeared, ranging from old Hall-Scott rail buses and streamlined McKeen cars to shop-built cars on big operations of companies like Georgia-Pacific, Weyerhaeuser, Rayonier, and others.

With the rise of organized labor and changes to the way loggers were paid, timber companies began to minimize transit time as much as possible. Instead of variations on piece work (the number of trees and size of trees felled in a shift), loggers were paid by the hour.

With that change, A Shay wending its slow way from the bunkhouses to the woods was romantic and picturesque, but the men on the cars were being paid an hourly wage to watch the scenery go by. Thus, the investment in a gas-powered speeder that could go twice as fast as the Shay would quickly pay for itself in increased productivity.

Some of these gas-powered crew vehicles were unique in terms of styling. If you're looking for modeling inspiration, some of the best photographs of these speeders/crew cars are in the books *Logging Railroads of the West* and the Labbe and Goe classic, *Railroads in the Woods*.

Link-and-pin couplers were still in use on the Louisiana Cypress in October 1950. Note the mangled grab iron in the foreground on the car at right. *C.W. Witbeck*

The Argent Lumber Co. also used link-and-pin couplers into the 1950s. At the mill, a worker looped a cable into a link to pull a string of cars to the unloader. *Jim Shaughnessy*

Available models

A quick survey of the Walthers catalog, although far from a complete listing of logging models, offers several items that logging modelers can readily use. The table on page 39 lists several models that might be appropriate for a logging layout.

There's a nice selection of equipment available in On2½, much of it manufactured in support of Bachmann's beautiful Shay. Although much of the rolling stock in this scale isn't based on specific prototypes, the cars are clearly "prototype inspired" and have captured much of the feel of the real thing. Most of this equipment is patterned after three-foot gauge prototypes, albeit shortened and slightly narrowed.

The listing doesn't attempt to cover the many craftsman and other short-run kits that have been produced. Magazines such as *Narrow Gauge and Shortline Gazette* and *The Timberbeast* are excellent resources for tracking down specialized models.

Track and turnouts

Logging railroad roadbed and track featured lighter construction than most common carrier lines. Logging lines were cheaply built, made to be both temporary and easily relocated. The coming of earth-moving equipment such as ditchers and steam shovels meant an improvement on the trunk lines of larger logging operations, but this didn't happen until the 1920s or later.

During the heyday of logging railroads, from 1890 through 1930, narrow gauge lines used 35- to 56-pound rail (measured in pounds per yard), with standard-gauge lines using 56- to 70-pound rail. In the Pacific Northwest, a few railroads that ran Mallets used 90-pound rail, which was used on many common carriers at the time.

Track is a scale model just like locomotives and cars, and for the

Efficiency demanded that log trains be long and heavy – often as heavy as the locomotive could handle. This late-1950s cupola view looks over a McCloud River train in California. *Donald Sims*

best effect, should be treated the same way.

The only commercial prefabricated track in O scale has rail that's simply too large to prototypically represent track used by real logging lines. The best way to get realistic track is by handlaying it. Code 70 rail (rail .070" tall) represents 35- to 40-pound rail, code 83 is 56- to 65-pound rail, and code 100 is 75-pound rail.

In HO scale, code 40 rail is roughly equivalent to 40-pound rail, code 55 represents 56- to 65-pound rail, and code 70 is 90- to 100-pound rail.

In HOn3, flextrack with code 40 and code 55 rail is available from Micro Engineering. Turnouts are available in code 55. All of the currently available HOn3 locomotives that would be appropriate for a logging line will operate well on code 40 rail. Older brass engines or models built from kits prior to 1960 might have deeper wheel flanges, meaning code 55 is about as small as you can go for trouble-free operation.

In HO, the same rule holds, but code 40 standard gauge track is no longer commercially offered.

Code 55 and code 70 track will handle anything except NEM (European) flanges and models produced between 1960 and 1965. Code 83 and code 100 track are too heavy to be prototypical.

In N, code 40 rail represents 90-pound rail, about as heavy as logging rail ever got, but it is the smallest rail available. Micro Engineering and others produce code 40 track but not turnouts. One compromise would be to use code 55 turnouts with code 40 flextrack. Real logging lines often used heavier rail in turnouts because it handled moving and general wear-and-tear better than light rail, reducing maintenance and improving reliability.

Your choices are limited in large scale, as cars and locomotives are made with oversize flanges that don't work on scale-sized rail, and the heavy track is designed for outdoor use.

The earliest logging railroads used ties that were little more than eight-inch-diameter logs with two sides cut parallel for stability. However, by 1900, having a neighboring sawmill with the capability of making ties meant that most logging railroads had nicely cut and shaped ties. Thus, commercial track, when coupled with the correct rail size, provides an accurate representation of logging track.

Trestles, bridges, and tunnels

Logging railroads followed the general lay of the land with minimal grading where the countryside was relatively flat. In hilly and mountainous country, surveyors attempted to smooth the grade to minimize stress on couplers and to avoid problems with slack running in and out as trains traversed went up and down. This meant that the track snaked along side hills, up canyon walls, and curved over streams (or stream beds) when it was necessary to move to the other side and in the opposite direction.

Logging railroads used trestles to maintain an even grade and to avoid earth fills. Although trestles require more long-term maintenance than fills, most logging branches were only used for a year or two so the added stability of a fill wasn't needed. Trestles were also good for drainage – they

This modern steel pulpwood car carries pulp logs from cutting areas to paper mills.

didn't block watersheds as with a fill. Trestles were susceptible to fire, and most had sand or water barrels placed on them, usually at every other bent (support).

Trestles on logging lines looked rough compared to those on mainline railroads. Trestles were never sloppily built – they were far too important – but the wood could be raw or rough in terms of finish. Early trestles used round logs for vertical posts. They were most likely cut nearby, trimmed of branches, barked, cut to length, and put into place with a pile driver or braced in a shallow pit.

When a pile driver was used, the log pilings were driven small-end first. The posts weren't likely to be the same diameter. The braces were sawn timber, but often of varying sizes and degrees of finish.

Logging railroads avoided steel bridges if possible, as they were expensive and difficult to build compared to pile trestles. Bridges were generally only used to span streams or rivers that were too deep for trestles, and in those cases the spans were kept as short as possible.

Tunnels were rare on logging lines. The few that existed were dug because there was either no other practical way to get to the timber or because the line was built with aspirations of becoming a common carrier.

In the early 1900s, larger operations such as Rayonier began to use steel trestles and bridges on their main lines from the loading areas. *Stan Kistler*

Pile trestles are signature structures of logging railroads. Most used untreated wood of varying dimensions, and most were designed to last only a few cutting seasons. *Darius Kinsey, Library of Congress*

Structures and logging equipment

A log train rolls through Rayonier Camp 14. The camp buildings at left are on the rails, ready to be moved to the next camp location.
Stan Kistler

When you think of a logging railroad, what images come to mind? Most modelers talk about the scenery, especially forests and mountains, but when pressed further, they will usually talk about a sawmill, log pond, spar trees, equipment in the woods, and logging camps. To a great extent, the structures and equipment associated with the forest products industry create the setting for our locomotives and cars. Although a forest can help set the locale of a layout, the buildings and associated equipment define the era and purpose as much as anything that rolls on the rails.

Logging railroads didn't have the same type of infrastructure as common-carrier railroads. Loggers had no stations, very little in the way of logos or standard colors for the equipment, and even the bridges and trestles were pretty basic.

FIVE

The Lombard Log Loader was a tracked vehicle for hauling logs and sleds on snow. It was steam powered and used a Shay engine.

Cutting trees in the woods

The first loggers in North America were here long before the Europeans arrived. Native Americans used the forest as a resource for building materials – the lodgepole pine was named because of its use in the structures of the Pacific Northwest. Harvesting techniques were simple – some trees were cut down with stone axes, but most were felled by controlled burning around the base until it could be pulled over.

The arrival of Europeans in North America immediately changed forest harvesting. Early settlers first felled trees to clear land for farms and to build houses and furniture. The axe was the primary tool for 17th century colonists, but saws – double-handled almost from the start – quickly came into use. Trees were felled close to where they were needed, with teams of horses or oxen moving them if needed.

The effort required to move a large log was enormous, and colonists who specialized in wood harvesting began using water to float logs whenever possible. In eastern Canada and New England, the use of water for transporting logs began well before the use of railroads. Many logs were cut in

The skidder operator hauls back the line from the road donkey to the yarding donkey (left). The fir log eventually arrives at the yarding donkey (right). *Two photos: Darius Kinsey, Library of Congress*

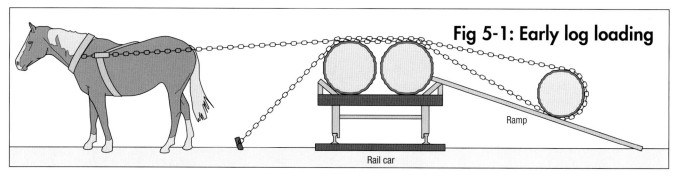

Fig 5-1: Early log loading

Ramp

Rail car

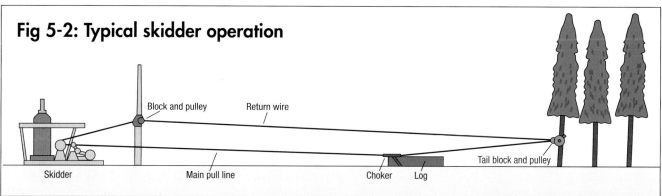

Fig 5-2: Typical skidder operation

Block and pulley

Return wire

Skidder

Main pull line

Choker

Log

Tail block and pulley

the summer and not moved until winter when the snow pack, combined with sled runners placed under logs, allowed easier movement.

The use of horse, oxen, and mule teams in the woods began early. Loggers learned to trim branches and cut a bevel on the bottom side of the butt of the log to reduce drag as the log was dragged through the forest. Even so, few logs were moved more than a thousand yards from where

they were felled to where they were hand sawn into timber.

By the end of the 1700s, as the trees nearest the coast were harvested and replaced with farms and buildings, the solid timber industry moved inland. The use of water power for sawmills began in the early 1800s, with the use of an artificial pond upstream to provide a constant supply of water for the mill's waterwheel. This pond also proved to be an ideal place to keep logs. They could be

moved more easily than on land, and they wouldn't catch fire. With proper mill design, logs could be moved by mechanical power – provided by the water wheel – to the saw blade, which was also powered by the wheel.

With this step, the modern sawmill was born. What was missing was an efficient way to move logs to the pond. The railroad would provide that answer, but with it came the need for better ways to cut the timber

The Hull-Oakes "wigwam" sawdust burner could be found at most Western sawmills through the mid-1900s. The conveyor carried sawdust to the burner from the vacuum blowers at the head rig. *Steve Austin*

Prior to 1900, teams of horses and oxen often pulled logs on iced sledways in the winter. This scene was photographed in Michigan around 1900. *Library of Congress*

and move it to the railroad.

A logging scene in the early 1800s would feature all animal power. Logs would be pulled by yoked teams of oxen, horses, or mules, with the forward end of each log beveled to provide a better shape for dragging. Beveling of logs continued as long as logs were pulled through the woods, even with giant steam donkeys and yarders.

The loggers in this early scene would all have double-bit axes (one edge used in the morning, the other in the afternoon) The loggers would also have double-handled saws, ropes, wooden block tackle for moving trees, and several wagons for moving their equipment. Living accommodations would be tents or very rough shelters with a cooking area. As time went on, accommodations improved but never lost their overall simplicity.

The loading operation usually

A team of 10 oxen skids a massive Douglas fir through a forest in Washington state around 1900. *Courtesy of the Weyerhaeuser Collection*

involved two teams of animals that pulled on parallel ropes running under the log, which was rolled up a ramp of several logs onto the log car **(fig. 5-1)**. This method was used in all areas of

the country until the era of the steam skidder and spar tree or the self-contained McGiffert and Barnhardt loaders. For small operations, this method of rolling logs up ramps and onto rail-

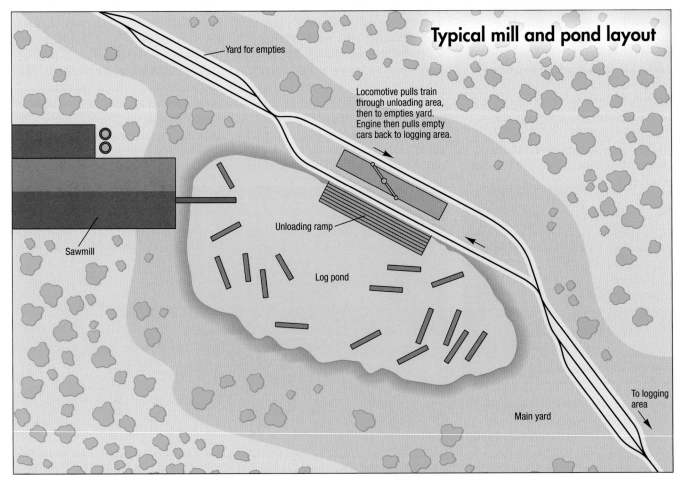

Typical mill and pond layout

Yard for empties

Locomotive pulls train through unloading area, then to empties yard. Engine then pulls empty cars back to logging area.

Unloading ramp

Sawmill

Log pond

To logging area

Main yard

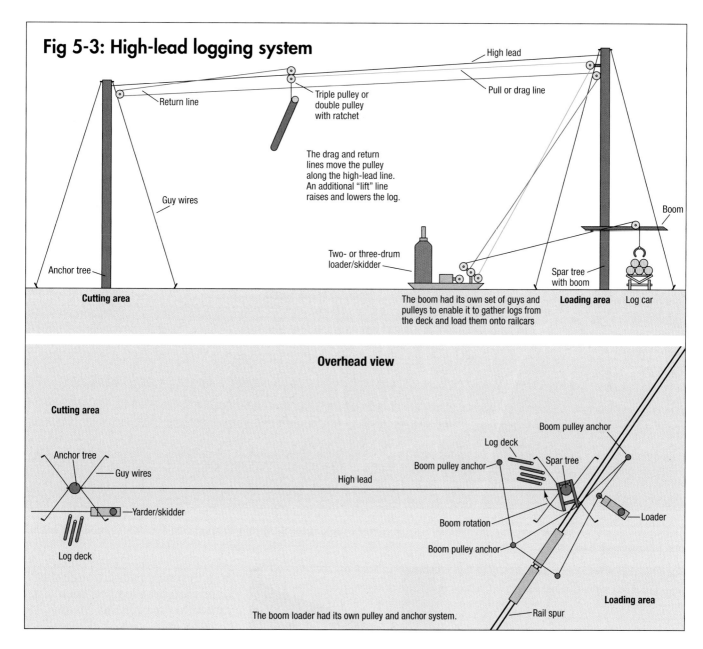

Fig 5-3: High-lead logging system

High lead

Return line

Triple pulley or
double pulley
with ratchet

Pull or drag line

The drag and return
lines move the pulley
along the high-lead line.
An additional "lift" line
raises and lowers the log.

Guy wires

Boom

Two- or three-drum
loader/skidder

Anchor tree

Spar tree
with boom

Cutting area

The boom had its own set of guys and
pulleys to enable it to gather logs from
the deck and load them onto railcars

Loading area   Log car

Overhead view

Cutting area

Anchor tree

Guy wires

Yarder/skidder

Log deck

High lead

Boom pulley anchor

Log deck

Boom pulley anchor

Spar tree

Boom rotation

Loader

Boom pulley anchor

Loading area

Rail spur

The boom loader had its own pulley and anchor system.

---

cars continued into the 1950s, albeit with a tractor or bulldozer instead of animals.

A late-1800s improvement to animal power were Michigan high wheels. These consisted of a large pair of wheels with a tongue for the draft animals and a heavy crossbar with a block for a wire rope or choker chain that would lift the lead end of a log off the ground. High wheels improved the ability of animal power to move heavy logs so well that they remained in service in many areas even after steam power came to the woods. Variations were used all over the United States.

By the late 19th century, the era of running logs down a river was over, but on the West Coast the movement of large rafts of logs via the sea was just beginning. This reached a peak in the 1920s and continued for some mills into the 1950s.

By the turn of the 20th century, most larger logging operations took advantage of steam power of some sort. The first of these steam-powered machines was the West Coast's famous Dolbeer Donkey. The Dolbeer and later varieties of steam donkeys were often referred to by their use, such as skidders, hoists, and yarders. In its most

basic form, a donkey comprised a vertical steam boiler powering a single-piston steam engine with a single-drum winch (all taken from merchant ship technology of the 1880s), all mounted on a substantial wooden skid.

At first, animals (horses, mules or oxen) were still used for local log moving and for loading logs onto railcars. Every single-drum winch used as a skidder required some method to get the choker (the loop of chain or cable that was put around the log to pull it) back to where the cut logs were waiting. Usually a horse or a mule was used (oxen were too slow) to

Few loggers could invest in a majestic steel structure like Simpson Lumber's Vance Creek Bridge in Washington. Built in 1929, it stood 347 feet tall and 827 feet long. *Chet Ullin*

The spar boom (here a double-beam crane) enabled one or two large steam skidders to log up to a one-mile radius around the spar tree. *Stan Kistler*

drag the cable back through the woods after the log had been yarded to the railhead. Steam donkeys and larger skidders all had whistles on them, and a complex code of whistle signals was used to warn loggers of movements.

An improvement was the two-drum yarder, which could haul a log in and then pull the choker back out to the woods **(fig. 5-2)** by having a block and pulley attached to a tree (the tail tree) in the area of the cut.

Logging's glory days

The glory years of logging, especially in the West, began about 1905 and continued until the start of the Depression. During that time, an area the size of Indiana was logged in California, Oregon,

Washington, Idaho, Montana, British Columbia, and Alaska. More than 5,000 miles of logging railroads were built and operated in this region, and nearly 2,000 steam locomotives designed for logging railroads were built during this time.

Technology matured as the small steam donkeys of the 1890s gave way to three-drum monsters that could pull a 50-foot-long, eight-foot-diameter log across a 1,500-yard canyon on a high lead line and drop it beside the tracks. With another drum working the loading boom, the donkey could lift the same log – while bringing in another on the high-lead line – and drop it gently onto a log bunk. When the car was full, a line attached to the lead car of the string could gently and precisely move the next car into place.

Logging had a strategic military importance during this era. During World War I, several railroads were built in the coastal mountains of Oregon to log giant spruce for making aircraft. So critical was this need that track and construction materials were floated ashore at remote beaches, allowing construction to start before connecting rail lines were built. These logging lines were known collectively as the Spruce Railroads, and they pioneered some technology that would become standard by the 1930s.

Among these advances was the change from hand- to gasoline-powered saws. The "misery whip," as the double-handled saw was sometimes called, was too difficult to use in some places. Gas-powered chain saws appeared by 1912, but first saw widespread use during World War I. The first ones were huge, requiring a team of men to handle, but by the end of the 1920s, they had become smaller and lighter.

Another development of that time was the Phoenix log hauler, which featured a steam boiler on a tracked chassis (like a bull-

The high-lead logging system used a spar tree to elevate a pulley so that the steel cable guiding the traversing pulley would be 50 to 100 feet off the ground. Another cable allowed a log to be raised and lowered from the pulley. *Darius Kinsey, Library of Congress*

When animal power was used to skid logs, the front end of the log was chamfered to keep it from digging into the ground. *Darius Kinsey, Library of Congress*

This steam loader has a steel tower that doubles as a skidder. In the foreground is a log on the high-lead wire coming into the loading area, and there's also a log on the boom of the loader. *Darius Kinsey, Library of Congress*

Louisiana Cypress elevated this steam loader so it could pull empties through and load them in front of the boom. *C.W. Witbeck*

Skidways were made of split logs laid in a V-shape on the ground. A small steam donkey is visible at right. *Library of Congress*

dozer). A two-cylinder Shay engine mounted alongside the boiler drove the tracks. They were eventually displaced by other tracked vehicles such as the Holt (later Caterpillar) tractor.

A Western logging line would run up a valley until it came to the logging areas. The tracks would climb to the ridges to the spar pole.

The spar pole was a tall tree that had been trimmed of all its branches and guy-wired to the ground for stability. It was the central point for the high-leads that went out to the cutting areas **(fig. 5-3)**. Most spar trees had the loader boom attached to their bases, but some had separate booms for multiple-track loading.

Companies such as Lidgerwood and Willamette made portable steel spar trees that could be carried on a special flatcar and raised at appropriate locations. These saved time and could be pre-rigged, enabling loggers to work a smaller area efficiently before moving on.

These were also the glory years of the logging camps – the towns that moved by rail from logging area to logging area. Bunkhouses, cookhouses, dining halls, medical clinics, and saw-filer shacks were built on skids so they could be transported on log bunks or specially built flatcars to the site of the next big cut. The big skidders and yard-

A giant redwood log leaves a trail of smoke as it goes down the skids. The skids were often greased to lessen friction and reduce the chance of fire. *Library of Congress*

The first step in making cedar shingles was cutting the logs into bolts, an arduous task at the turn of the 20th century. *Darius Kinsey, Library of Congress*

ers, which could move themselves through the forests as needed, winched the buildings onto the cars and then moved themselves into place. The book *Railroads in the Woods* by Labbe and Goe has some excellent photographs of a log camp on the move. A camp might move once or twice a season; loggers who worked year-round might move five or six times a year.

Depression and World War II

The golden era of logging ended with the Depression. By 1932, more than half the logging operations in Oregon and Washington were idle. Some remained that way until the end of the decade, and others were abandoned in place. Some lines were kept alive, but barely. Mills, running short hours with skeleton crews, cut logs that had been hauled in by trucks. Small at first, these trucks were precursors of today's giants that vanquished the railroad and its inherent inflexibility.

The few logging railroads that stayed in operation through the Depression modernized. The Pacific Coast Shay, arguably Lima's finest product, along with competitors from Heisler and Climax, replaced dozens of smaller, older locomotives. The reload line became as common as logging lines themselves, and the industry began moving toward a future in which logging was a carefully engineered business.

World War II brought a much-needed boost to the logging industry. The construction boom during and after the war helped keep logging railroads going well into the 1950s. The last non-reload logging line shut down in 1957 at Keasey, Ore., when Long-Bell (a division of International Paper) closed its last cut using a high-lead to load log cars.

After that, logging railroads were strictly reload lines, and those remaining underwent gradual dieselization or eventual

Several men pose near a steam donkey at a high-lead logging operation in the Pacific Northwest around 1920. The cables at right go to the spar tree and the high-lead line that runs to the cutting area. *Courtesy of the Weyerhaeuser Collection*

A lumberjack gathers logs together to assemble a boom, or large raft, to float down Washington's Skagit River in 1906. *Darius Kinsey, Library of Congress*

Once the logs have been floated down the river, the green chain (or lifting chain) pulls logs from the river up a ramp and into the mill. *Library of Congress*

When the side stakes are released, logs tumble from skeleton cars into the pond at MacMillian-Bloedel's Chemainus, British Columbia, mill. *Clinton Jones, Jr.*

This view shows how the unloading trestle extends across (instead of next to) the pond at MacMillian-Bloedel's Chemainus mill. *Clinton Jones, Jr.*

Louisiana Cypress used an overhead wire unloader for its mill pond at Harvey. *C.W. Witbeck*

abandonment. Trucks could work small cuts more efficiently, and cutting was mechanized. Tracked vehicles with hydraulic booms could fell, de-limb, buck, and load a log in less time that it used to take to set the choker.

Although the style of logging had changed from the glory years, this is still a fascinating era to model. The trains look the same for the most part, with geared or rod locomotives bringing strings of loaded log cars to the mill and taking empties back to the reload site. Big trucks, Caterpillar tractors, skidders, and other fascinating equipment abound. Most of these operations were on a fixed railroad plant that hadn't changed much since the 1920s.

By the 1970s and '80s, what was left was either modernized or had become a ghost town. But new and rehabilitated lines now haul wood chips, pulpwood, and forest products over rails that may once have hosted Shays and Mallets. First-generation diesels pull solid freight trains of lumber or paper from mills to junctions with the larger world of common carriers.

The company mill

The sawmill is the heart of any logging railroad. Cut trees have to be processed, and the sawmill was where trees became boards. Early sawmills (water-powered ones especially) used reciprocating-blade saws, usually ganged in twos or threes to enable multiple simultaneous cuts. These early water-powered mills could saw a 16-foot-long, two- or three-foot diameter log in 20 to 30 minutes if the blades were sharp and there was plenty of water. About 12 logs could be cut per shift, and in the days before electric lights the mills tended to operate only during daylight hours.

The next change was the move to circular saws and steam power. The mill could then be located much closer to the logging site, and with the added power of

This 1997 view of the Hull-Oakes sawmill in Monroe, Ore., shows the layout of the mill buildings and pond. *Historic American Engineering Record*

steam, logs could be cut more quickly. Powered log handling through the saw, the invention of the power cutoff saw, and other improvements came slowly during the steam era. The circular saw cut faster but its diameter limited the size of log it could handle, and it couldn't be ganged for multiple cuts on a single pass as with a reciprocating saw.

Small steam sawmills were sold through Sears and other catalog companies, aimed at farmers and small operators. A small steam sawmill was in operation along the Denver, South Park & Pacific as early as 1878, providing construction timbers and boards for the railroad and the mines in the area. At one time, Keystone made an excellent model of a small, single-blade circular saw mill in O scale.

Electricity provided the key to increasing production, and by the early 1900s most large sawmills were powered with electricity. These mills had multiple band saws with powered log carriages and could process a four-foot-diameter, 24-foot-long log in

under 10 minutes. It was this change that made the large, high-capacity sawmill the focal point for large logging railroad systems.

These mills could be found wherever there were large forests being harvested. The need to provide large numbers of logs every day led to the high-capacity locomotives and large-scale logging operations of the glory years of logging.

A sawmill complex of this era had four basic elements: the sawmill building with connected structures, the mill pond, the tracks from the railroad to the pond, and the drying yards and drying kilns. The mill pond's size and shape varied with the location of the mill, but it was usually 10 to 20 feet deep and large enough to hold a two-day supply of logs at a time. The pond area sometimes included a log storage area called the cold deck where logs could be stored until needed – usually during seasons when the woods couldn't be logged.

A fixture near most mills, from about 1900 until the 1950s was a

large, usually rusty, conical sawdust burner, often called a wig-wam burner. About 15 percent of every log is turned into sawdust, and another 15 to 20 percent is unusable as lumber. The off-cuts could be burned for fuel in the powerhouse boilers, but the sawdust was disposed of in the burner.

These burners disappeared in the 1960s as sawdust became useable in various products, and as new air pollution regulations were enacted.

The mill pond
From the late 19th century until the 1960s, most sawmills had ponds for handling logs. From the pond, logs were fed to the green chain (or lifting chain). This was a continuous conveyer chain with hooks and sidewalls that carried logs up from the pond into the mill. Logs, even giant ones, could be handled with small motorboats rather easily. Logs in the water wouldn't dry out, crack, or catch on fire.

The mill pond performed the same function for the sawmill as a

The Argent Lumber Co. employed this end-boom loader in South Carolina. *Matt Coleman collection*

Argent's three-drum yarder was out of service by the time of this 1966 photo. *Matt Coleman collection*

classification yard did for a railroad. It allowed logs to be sorted by size, species, and length. Sawmills filled orders for wood of specific types and dimensions, so the ability to easily sort logs was crucial to efficient operation.

Although ponds were the method of choice for most large mills, they did have problems. Ponds freeze, making winter log movement impossible in some climates. Logs sometimes sink, so mills had to regularly drain ponds to extract dozens of logs that had quietly disappeared from the surface (a situation more common with hardwoods than softwoods). Ponds also require regular maintenance and a source of water flow.

Model sawmill ponds are usually more representational than scale. Real log ponds were normally five to 10 times the area of the sawmill. Thus, a 200 x 300-foot sawmill (6,000 square feet) would need a 60,000-square-foot pond to hold logs and provide room for sorting them. This would overwhelm most logging layouts, so most modelers simply model the portion of the pond nearest the lifting chain.

Sawmill ponds were almost always made by damming a stream or river and building the mill and railroad alongside. Some operations, especially in more hilly areas, dammed a stream and then diverted the water to a more convenient location where a pond was constructed in a low area.

Ponds had log unloading areas where trains of loaded logs were pulled alongside the pond. A mechanical device would then push the logs off the cars and down a ramp into the pond.

The track along the unloading area usually had guardrails, because log cars tended to derail if the logs became hung up on the log bunks. The ties were usually covered because of the huge amount of bark that was knocked off during unloading.

Actual log unloading would be difficult, if not impossible, to duplicate in a model, although large-scale modelers working outdoors might give it a try.

Thanks to today's smaller logs and larger log-handling equipment, modern sawmills don't have log ponds. Instead, trucks and railcars are usually unloaded by large forklifts, and the logs are stacked and pulled from a cold deck. The cold deck is often protected with a water spray to keep logs wet and protected from fire and wood pests.

Sawmill track arrangements

The track layout and equipment operation are what make sawmill trackage so fascinating to model.

This consists of the actual log-unloading trackage and the associated track needed to get the locomotive back to the other end of the cars, along with the often-extensive shop with storage tracks for the equipment.

Because logging and sawmill operations were self-supporting, the size of the shops and locomotive servicing area would be quite large compared with conventional short lines of equivalent size. During the glory days of logging, larger operations had an engine house and a car shop with an extensive woodworking area, and some even had a boiler repair shop to handle locomotives as well as steam boilers for the woods equipment.

For narrow-gauge loggers, some sort of transfer track would be needed for when a loco was sent off line for repair or rebuilding. Otherwise, a few interchange tracks would suffice. On the opposite side of the mill, though, there would be extensive trackage for loading finished lumber into boxcars and flatcars. Often the switcher in the finished lumber/shipping area was an old logging locomotive now too small to be efficient, or an old rod locomotive from a mainline railroad.

The drawing on page 52 shows a generic track layout for a small- to medium-size sawmill. Note how the track is arranged so the locomotive can pull the cars past the log dump, then uncouple and run around the string of cars to pull them back into the yard ready for the next turn to the woods. Loggers avoided pushing strings of log cars, whether loaded or empty, for long distances. Historical notes from the lines that now make up the Cass Scenic Railroad show that most derailments happened on switchbacks and at loading sites where cars had to be shoved into place.

Drying yard and kiln

Once lumber was cut, it needed to dry before being shipped to customers. Air drying was most common in early mills, but kilns, with their even heat, reduced the amount of time between sawing and finishing as well as the time until the lumber was shipped.

The tracks among the drying yards and kiln could almost amount to a separate railroad (and for some mills, it was a separate narrow-gauge network of track). Stacks of cut lumber were put on flatcars, some specially made for this service, and moved to a drying stack. Here lumber was stacked with spacers between layers so that air could flow around the lumber to remove moisture. When a kiln was used, the lumber might be stacked on special cars that could be put directly into the kiln. For high-value cut lumber, such as hardwoods destined for furniture makers, covered storage for dried wood was a necessity.

The entire kiln, drying yard, and finished-lumber shipping operation was usually on the opposite side of the mill from the logging railroad side. In a nutshell, logs come in one side of the mill and are moved through the pond into the sawmill. Cut lumber came out the other side, was dried, placed into storage, and then moved to the final shipping area.

The complexity of a large sawmill is what makes it so

fascinating. To accurately model all areas of a mill would be a major undertaking but one that could be the cornerstone of a very comprehensive layout. For those with the space and interest, a large layout with a full paper mill and sawmill could keep several operators busy with just the forest railway and mill switching operations, not to mention mainline operations for both the industries.

Paper mills

Paper mills, especially the larger ones built after the late 1920s, were major consumers of wood, fuel (either coal or oil), and process chemicals. They are major shippers of both paper and wood chemicals recovered from the pulping process.

A large paper mill, one that produces more than 1,000 tons per day of paper, requires more than 2,000 tons of wood every day. (The exact amount varies by type of paper and, for modern plans, the amount of recycled paper used.) Most period pulpwood flatcars could carry about 40 to 45 tons of wood, so 40 to 45 carloads of wood would need to arrive every day and the empties taken out. Today's large wood chip gondolas carry up to 100 tons, and the mills they serve produce from 1,500 to 2,000 tons of paper per day.

Paper mills that burned oil to supplement the firing in the recovery boiler and lime kiln required at least one tank car of fuel oil every three days. Mills pulping pine or other softwoods would generate a tank car load of turpentine per week and a tank car load of tall oil (an oil from pine resin) every 10 days. And this doesn't count inbound carloads of caustic soda needed for the chemical processes. Such a mill would generate at least 20 to 25 boxcar loads of paper per day, unless some was shipped by truck.

Paper mills are often located near sawmills, sometimes as part of the same complex, especially in

This Barnhart loader could pull empty cars through itself from behind, allowing it to load logs on multiple cars without a locomotive.

the South. Today's large, modern paper mills are a product of the late glory years of logging. Before that period, papermaking was a more localized industry with smaller plants.

Since the paper industry covers the transition from the 1940s through the 1960s when steam logging ended, a paper mill can still be a reason for a forest railway on a modern layout.

Wood chemicals

The wood chemicals industry flourished in New York and Pennsylvania from about 1870 to 1910, using the distillation method of turpentine recovery to produce a variety of wood chemicals as well as charcoal.

The modern charcoal briquettes we use in home barbecues are made by this same method of closed-chamber destructive distillation of hardwoods. The wood is placed in a sealed chamber that can be heated, and turpentine and other volatile compounds are driven off and recovered with a special distilling apparatus. The remaining wood is turned into charcoal and compressed into briquettes.

Many of these early wood chemical plants are discussed in *The Logging Railroad Era of Lumbering in Pennsylvania*, a 13-book series by Benjamin Kline, Jr., Walter Casler, and Thomas Taber III. Many of these operations were served by small logging lines using an 0-4-0T or other small engines to haul a few carloads of wood each day from the forest to the plant.

If you're seeking a smaller logging-related industry for a model layout, this might be the ticket to a unique, interesting-to-operate layout.

Available models

The number and variety of good logging-related models is overwhelming in HO, and nearly so in N scale. Among the best are Walthers' sawmill and paper mill series from several years ago and the BTS logging series, which is stunning in accuracy and thoroughness.

Each prototype sawmill was unique, although most followed a similar basic layout. Most sawmill and lumber mill buildings are made of wood (except for the sawmill's boiler house and the sawdust burner).

A sawmill, like other manufacturing endeavors, is a collection of separate processes and functions, many of which are housed in a separate structure or a location near the actual mill building. Each one of these is important and should be included or somehow indicated.

The lumber drying and shipping areas were typically next to the tracks, with tracks also alongside the log pond or deck. The mill building itself tended to be set back from the track, perhaps to avoid cinders from locomotives.

Several manufacturers offer sawmills and related structures, including many laser-cut wood kits. Be sure that the models you choose are suitable for the era you

The logs have been sorted and bunched at one end of the pond at the Southwest Forest Industries mill in Flagstaff, Ariz., in this mid-1960s view. *Eric Lundberg*

This A-frame unloader at Hull-Oakes, serving trucks in the late '90s, is typical of those used in many areas for unloading rail cars into the mill pond. *Historic American Engineering Record*

are modeling and will fit your available space. Also, factor in the other necessary elements such as the pond, cold deck, lumber storage area, and sawdust burner.

Auxiliary equipment, such as bulldozers, haulers, cranes, steam donkeys, loaders, forklifts, and trucks, should be matched to prototype photos and modified accordingly. The Walthers catalog lists hundreds of suitable vehicles, and Rio Grande Models makes some detailed, highly accurate models of logging equipment.

Vehicles and equipment for moving cut lumber around a mill

are necessary even for the smallest of mills. By the 1890s, most sawmills had some sort of trackage (usually narrow gauge) with cars or carts to move stacks of cut lumber. In HO, several small rail push cars designed for track maintenance could be modified for use on this very light rail (handlaid if possible). Later eras might use specialized forklifts for the same operation.

The more you understand how a log was transformed into finished lumber, the more accurately you can model logging areas and sawmills.

Designing a logging layout

The subject of layout planning has inspired countless articles and many books over the years, but I've seen very few plans that effectively capture the track layout and atmosphere of a true prototype logging operation. Designing a logging layout requires a thorough understanding of the purpose and limitations of handling large logs in the middle of a forest, both on the ground and on a railroad. Coupled with the information found throughout this book, this chapter is designed to help modelers create a logging layout based on prototype practices. A realistically planned layout will enable prototype operations and also show cars, locomotives, and other equipment at their best.

This 1950s overview of the West Side Lumber Co. car shops in Tuolumne, Calif., shows the dual-gauge track with the mill switcher – a former narrow-gauge Heisler that was converted to standard gauge. *Glenn W. Beier*

This diminutive GE 44-tonner is typical of diesel power often found on post-steam-era forest products lines. Models of small locomotives handle tight curves well.

Logging railroad design

For starters, look closely at what makes logging operations different than operations on regular common-carrier railroads. These differences will affect layout planning and design.

Logs are both a modeling problem and the reason for an interesting layout. A true logging layout focuses on the continuous cyclical operation of loaded trains coming to the sawmill (or sawmill trackage) from the woods, with empty log cars returning from the sawmill to the cutting site. You have to somehow be able to make logs appear to move on and off of the log cars without resorting to toy-like animation.

Before we delve too deeply into the design of a logging model railroad, let's look at the four elements of modeling logging railroads discussed in previous chapters: era, location, type of operation, and scale and gauge. Each of these will have a major impact on the type of layout you can design and build in a given space. Many modelers are already committed to a scale and gauge, determined by existing equipment, an existing layout, or budget limitations that preclude certain sizes of equipment.

We've all seen model railroad layouts that are well designed and superbly executed, but lack coherence in theme because the builder has too many interests to be adapted to a single layout.

One in particular comes to mind – the primary layout was based on the New York Central of the 1960s, just prior to the Penn Central merger. But because of an interest both in narrow gauge and logging, the builder had incorporated a rather nice narrow-gauge branch line that looked very much like the Denver & Rio Grande Western's Silverton branch in Colorado. On another section of the layout he had built a logging operation, complete with nicely modeled pine trees (and lots of them), that was clearly based on the West Side Lumber Co.

As beautiful as each of these layout elements were in their own right, they were somewhat discordant as a whole. Taking nothing away from the effort, I wondered if his interests might have been better served by having a separate but small logging branch in Sn3, On3 or even Gn3 where his love of building detailed logging equipment might have been better exhibited – and perhaps the same for his narrow gauge line. With that thought in mind, some of my suggestions for layouts will include building a second (or even third) layout in a different scale to meet the interests you might have for adding a logging operation to your model railroad.

Location, location, location

Real estate professionals will tell you that the value of a property is

Lumberjacks work to roll a log up a ramp to top off a car on this great north woods line. This would make a great mini-scene on a model railroad. *Library of Congress*

Box mills such as this one were common in California and other agricultural regions. Specialized saws and cutters made the thin boards that were manufactured into orange crates and other containers. *Library of Congress*

in its location more than in the architectural amenities or style. Although the analogy isn't perfect for a model railroad, there is some truth in it.

Layouts are often designed for reasons of equipment availability, beauty of a prototype location, or having information about a certain logging line. The tremendous number of layouts based on the West Side Lumber Co. operation in Tuolumne County, California, are not due as much to modelers having any particular love of that area (although it is indeed beautiful) but because there have been four major books written on this railroad and nearly every locomotive that has run on its rails has been offered in HOn3 and On3. Not that this is bad, but the tremendous variety and interesting locations for logging railroads are such that to be limited to only one specific railroad in one location puts a damper on many for participating in this area of the hobby.

The increasing number of excellent and moderately priced logging locomotives (the Bachmann three-truck Shay and two-truck Climax, the Roundhouse two- and three-truck Shays in HO and Hon3, and the venerable but wonderful Rivarossi two- and three-truck Heisler) allow modelers to move much closer to home, if they want, to re-create a logging operation.

The variety of brass locomotive models imported in the 1960s, '70s, and '80s, although more pricey than plastic models, provide modelers with an even broader selection of location and era. Also, the variety of rod locomotives, especially in HO standard gauge and now in On30, allow smaller and more obscure lines to be modeled with far greater fidelity than was possible before.

Choosing a locale

Every region of the country except the Great Plains (and the state of Rhode Island) has had more than a few logging lines, even if only for a season or two. Big-time railroad logging got its start in the pine forests of Michigan. It was in this state, a moderately hilly part of the country, that the need for something other than a rod-connected locomotive became the driving force for Ephraim Shay to develop the best-known logging locomotive of North America.

The soft, swampy ground found in much of the woods of Michigan gave rise to the famous Michigan High Wheel log carrier, which allowed animal power to pull large logs out of the forest to the railhead or the sawmill. All of this came just as the concept of the pure logging railroad was beginning to mature. Modelers choosing this area have their choice of everything from Civil War-era locomotives and ancient flatcars to modern Shays and steel log bunks, depending on era, and can run them on track that twists and winds among huge trees.

Except for the tree species, modelers choosing Appalachian railroads from West Virginia to northern Georgia will find many similarities to Western logging. Three-truck Shays, Climaxes, and Heislers abounded in this region; many were coal-fired, although wood (and even some oil-fired) engines were used through the final days of steam operation.

Steep hillsides, granite outcroppings, and a mixture of hardwood and softwood would make a layout set in the Great Smoky Mountains potentially stunning in appearance. The 2007 issue of *Model Railroad Planning* included an excellent article on modeling this type of forest. And like the West, the use of large

This unnamed logging town in Washington is typical of those found in the Northwest at the turn of the century, with buildings clustered along the river near the sawmill in the foreground and tall trees rising up a steep hillside. *Darius Kinsey, Library of Congress*

Mixed Train Daily – covers some of these logging lines.

The era determines the effort

The farther back in time that a layout is set, the more effort it will take to model it, mostly as a result of limited information and equipment. The majority of books (and available models) cover logging and forest product lines from the 1940s to the present. Some books include photos of pre-1900 logging, but these are rare, and few commercial models can be adapted or modified to accurately capture the look of these early prototypes.

Those who want to model the early years of logging face some challenges. For one thing, much of the character of early logging railroads comes from the light – almost delicate – locomotives and rolling stock. In HO, early versions of Climax, Dunkirk, and Shay locomotives have been offered in brass. A few rod locomotives exist, including some early 4-4-0s in brass (including the *William Mason* and the Promontory ceremony locomotives from the Central Pacific and Union Pacific).

Several brass 2-4-4-2 models have been imported in O scale, as well as a large number of geared locomotives in standard as well as On3 and On2, making it feasible to model logging in O standard or narrow gauge.

Chapter 2 provides additional guidelines for what will and won't work for locomotives.

Of course scale-height rail is mandatory for accurately modeling early railroads. In the 1880s, the heaviest rail in common use on mainline railroads was only 68 pounds per yard – the equivalent of code 60 rail in HO and code 100 in O scale. Since logging lines weren't built with new rail to mainline standards, plan on code 55 in HO as the maximum, or code 40 if you are willing to lay your own track.

non-locomotive steam power on skidders, two- and three-drum yarders, hoists, and other steam loaders offers lots of modeling potential.

Logging operations in the Mid-Atlantic and Northeast were mainly small, and some railroads worked the same locales for many years. *The Logging Railroad Era of Lumbering in Pennsylvania*, a 13-book series by Benjamin Kline, Jr., Walter Casler, and Thomas Taber III, is the bible of eastern logging. It covers everything from the hemlock harvests of the central Appalachian region to the hardwood trees harvested for the

charcoal and wood chemical industry of Pennsylvania and New York.

The mid-Atlantic and southeastern coastal regions from Maryland all the way around to Texas are fertile territory for modeling. These areas, along with the deep piney that stretched inland for hundreds of miles in some areas, featured logging railroads that were rarely seen by railfans, even in later years. Mallory Hope Farrell's recent books *Slow Trains Down South* (volumes 1 and 2) – an update and expansion of Beebe and Clegg's seminal masterpiece

This turn-of-the-century logging camp is in Michigan, but it shares a look in common with many camps throughout the country. *Library of Congress*

The ease of using scale rail for modeling early logging lines is an advantage in O scale. In O scale, code 70 represents prototype 35-pound rail and code 83 is 55- to 60-pound rail. Even code 100 rail is big enough to run logging Mallets, but still light enough to capture the essence of a logging line.

For modelers with the interest and the skill, early-era logging layouts are naturals for fine scale (Proto:87 in HO and Proto:48 in O scale) modeling. These standards specify wheels with scale-size profiles and flanges and track with scale-size flangeways and components for improved appearance. See www.proto87.org and www.proto48.org for details.

Limitations on design

A potential problem is that logging railroads don't offer the normal switching operations of a common carrier, where locomotives have to sort individual cars from trains and place them in specific spots at industries. Most conventional layouts have yards and other tracks allowing this to be done.

Instead, logging railroads are, in effect, unit-train operations. Photographs of logging lines in their later years often show the same blocks of cars, in the same order, over periods of several weeks. Strings of log cars don't have to be broken apart and shuffled around. This is why many people who model logging turn to unprototypical additions such as mixed trains to get more variety on a railroad.

However, doing this makes a logging line less focused than it should be. Although adding non-prototypical operations might make the layout more interesting to the builder, it will often confuse viewers. Instead, the best way to adapt prototype logging operation to a layout is to treat a group of cars as an operating unit and not as single cars.

Workers put the finishing touches on a tall timber trestle on Washington's English Logging Co. line. *Library of Congress*

Three key elements of a logging layout are displayed in this photo of a Weyerhaeuser line: a geared locomotive, tall wood trestle, and loaded log cars. The boxcar is likely bringing logging supplies to a camp. *Photo courtesy Weyerhaeuser Historical Collection*

Types of layouts

There are many factors to consider when designing a layout, including size, style, prototype, operations, and scenery. A great general guide to layout design is John Armstrong's classic book, *Track Planning for Realistic Operation* (published by Kalmbach).

Start by considering the space you have available. A 4 x 8-foot (or equivalent) layout is considered small, but provides room for a decent amount of track in HO or On3. Building a small layout gives you the chance to have it up and running in a reasonable amount of time.

Room-size layouts (around 12 x 12 feet) offer more flexibility. Building an around-the-walls layout in such a space allows broader curves, better separation of scenes, and a longer run of track. Model railroads this size can still be built in a reasonable amount of time and be maintained by a single modeler, and can be located in a spare bedroom or part of a basement.

You'll often see larger layouts, including some huge basement empires, in the pages of *Model Railroader* and other hobby magazines. This extra space is generally used for longer main lines and increased visual impact

Small geared steam locomotives, such as West Side Lumber Co. two-truck Heisler No. 3, are ideal for model railroads. *Glenn W. Beier*

of the trains as much as it is to increase the operational intensity.

In addition to the size of a layout, the type of layout is also important. There are several basic layout themes to consider, the first being operational intensity.

Some layouts pack a lot of track – such as switchbacks, crossovers,

crossings, and multiple sidings – all in a small space. These operationally intense layouts aren't readily adaptable to logging layouts because the type of operations don't replicate what happens on a true logging line.

Another type is the scenery-intense layout. Many larger pikes

Loggers often rode between camps and logging sites on ordinary logging cars. *Photo courtesy Weyerhaeuser Historical Collection*

Tight clearances are common around mill buildings, making small locomotives necessary for switching. *Glenn W. Beier*

Mill ponds were not tranquil swimming holes, as witnessed by the bark and debris at the edge of the Louisiana Cypress mill pond in the 1950s. Ponds were quite large, making them candidates for selective compression on model railroads. *C.W. Witbeck*

fall into this category, but some 4 x 8-foot layouts qualify as well. These generally feature a loop of track, sidings for a train or two, and scenes filled with scratchbuilt structures and intensely detailed areas. For modelers who love logging equipment and structures, this approach is well worth exploring.

A third type is what I refer to as "equipment intense." As an example, for a lover of locomotives, the shop area of a large logging operation would be fascinating to model. It could consist of a shed, shops, and all the support facilities needed for steam locomotives. Potential models include an enginehouse with large doors to allow viewing the interior, an outdoor track for locomotive storage, and facilities for repairing locomotives, log cars, and other equipment such as the yarders, towers, and skidders that would be brought in for maintenance.

The final layout type would be a captive common-carrier short line serving a sawmill or paper mill. Such a layout could combine interesting logging equipment and structures with loose-car railroading. This offers the potential for an exciting layout, and allows

running equipment that is about one generation older than what is being used on the connecting mainline railroad.

Prototype vs. freelance
Most prototype logging lines had signature elements that were unique to that particular railroad.

These items can include locomotives, structures at a specific location, scenes, and log cars and other rolling stock.

Certain prototype logging lines (the West Side Lumber Co., for example) have had so much broad exposure in various publications over the years that knowledge of

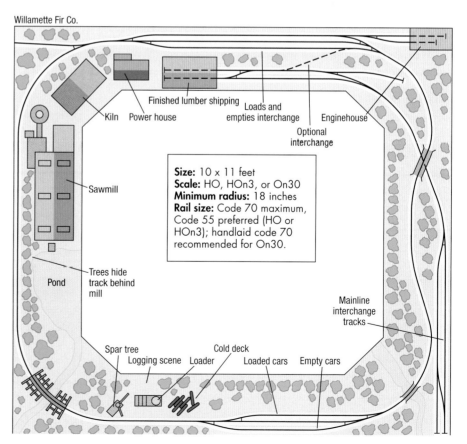

Willamette Fir Co.

Kiln Power house Finished lumber shipping Loads and empties interchange Enginehouse Optional interchange

Sawmill

Pond Trees hide track behind mill

Size: 10 x 11 feet
Scale: HO, HOn3, or On30
Minimum radius: 18 inches
Rail size: Code 70 maximum, Code 55 preferred (HO or HOn3); handlaid code 70 recommended for On30.

Mainline interchange tracks

Spar tree Logging scene Loader Cold deck Loaded cars Empty cars

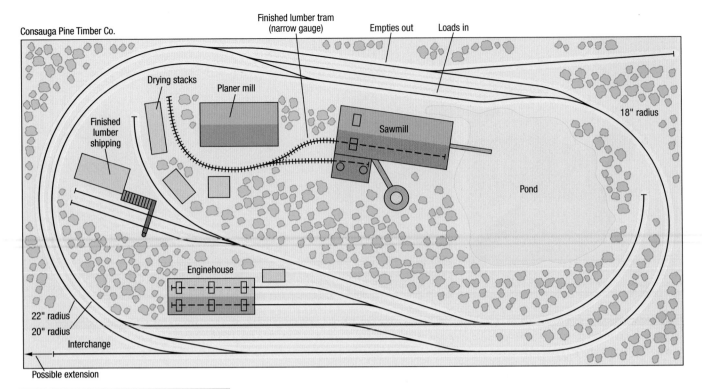

Consauga Pine Timber Co.

Finished lumber tram (narrow gauge)

Empties out

Loads in

Drying stacks

Planer mill

Sawmill

18" radius

Finished lumber shipping

Pond

Enginehouse

22" radius
20" radius

Interchange

Possible extension

Size: 4 x 8 feet
Base: Plywood or foam sheet
Scale: HO, HOn3, or HOn30
Minimum Radius: 18 inches
Rail Size: Code 70 maximum, code 55 preferred.

them is almost universal in the modeling and logging fraternities. They have also been well-represented in the number of models based on their prototypes.

Using these signature items on a layout can positively reinforce

the fact that you're modeling a specific prototype. The same items, however, can become quite distracting if you use them on a freelance layout.

If something stands out as being from a particular railroad, no matter how you paint it, even if you have a plausible explanation for it, a knowledgeable viewer will spot it immediately, and it will detract from the realistic effect that you're trying to create.

If you are going to freelance, stick with generic locomotives, cars, and buildings, and truly develop your own fictional line.

You still have many options open to you in freelancing. For

Size: 4 x 8 feet
Scale: HO
Base: Sheet extruded foam
Minimum Radius: 18 inches
Rail Size: Code 70 maximum, code 55 preferred (For commercial track, use code 70 turnouts and code 55 flextrack).

Snoqualmie Lumber Co.

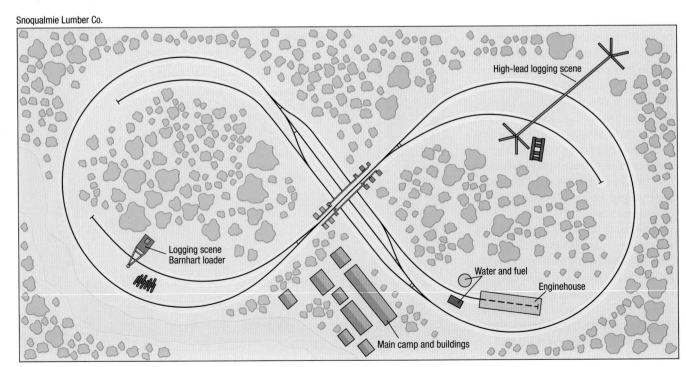

High-lead logging scene

Logging scene
Barnhart loader

Water and fuel

Enginehouse

Main camp and buildings

Size: 5 x 8 feet (approx.)
Scale: HO
Base: Plywood
Minimum Radius: 18 inches
Rail Size: Code 70 maximum, code 55 preferred (For commercial track, use code 70 turnouts and code 55 flextrack).

example, most of the major builders of geared and rod locomotives had stock catalog locomotives for sale. Bachmann's HO scale three-truck Shay is a beautiful example of a locomotive that could be used in the East, West, or South with plausibility. Your freelance line may have also purchased an engine second-hand.

Focusing on a specific prototype, no matter how obscure or common, will keep you on track toward building a realistic layout.

Let's take a look at a few small and medium-size track plans for logging layouts.

The Willamette Fir Co.

The focus of this around-the-walls layout is on the logging equipment and the sawmill complex. This layout plan contains everything from the log loading area to the sawmill. I have set the HO version in the glory years of the 1920s in the west side of the Willamette River Valley in Oregon, in the low rolling hills that were once home to magnificent stands of old-growth Douglas fir.

Appropriate locomotives include the Bachmann Shay and Climax and the Rivarossi three-truck Heisler.

The opportunity to build highly detailed, accurate mill structures can offset the somewhat limited operational opportunities that are inherent in a pure logging line. The sawmill complex, if set in the 1920s, could be something as detailed as the BTS craftsman wood kit for the McCabe Sawmill, which includes a sawdust burner, drying kiln, loading shed, planer building, and some outdoor drying stacks. It could also be simplified using injection-molded plastic kits.

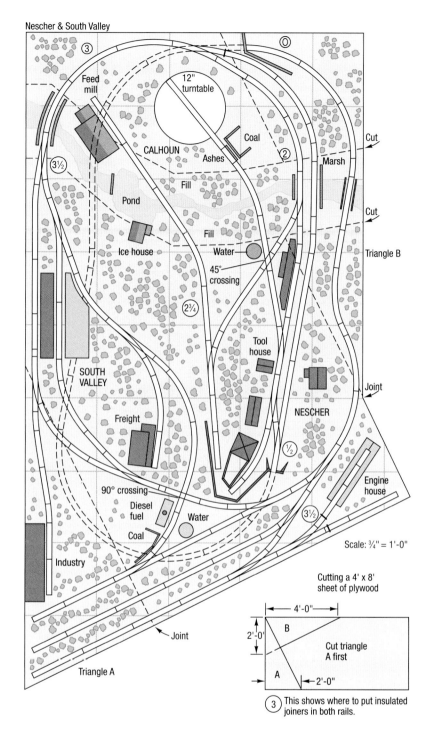

If you want a more transition-era setting, consider looking for some of the sawmill structures made by Walthers in the 1990s. These were beautifully done in plastic and would move the era forward into the 1940s or '50s.

The layout could also be built in On3 or On2½ with minimal modifications.

Operation on this layout would ideally use two locomotives. One would take a train of empty cars

out of the yard by the mill and cross over the mainline branch to the shipping shed and then to the spar tree and loading area. Here, a runaround is set so the locomotive can move around the cars and then gently push them beyond the skidder and spar tree boom loader. On a real railroad, this track would have been built on a slight upgrade so the cars could have been eased down one at a time, with a logger riding the

Fire was a constant threat around logging railroads (especially steam-powered lines) and sawmills. Note the fire hydrant next to the locomotive cab, and the water tank car for a firefighting train at right. This is Southern California's Pickering Lumber Corp. *Matt Coleman collection*

If your layout doesn't have room for a large trestle, use a small one. This Shay is crossing an Appalachian stream. *Matt Coleman collection*

It was rare for post-World War II logging lines to lay new track to cutting areas, but Louisiana Cypress was still doing it in 1950. Note the light rail and lack of tie plates. *Two photos: C.W. Witbeck*

brakes, and positioned under the loading boom. When the cars are loaded, the locomotive returns and hauls them back to the mill.

On this layout, the empties are pushed onto the track hidden by trees behind the mill, and the train of loaded cars would have been pulled by the second locomotive behind the mill (nominally for a trip past the pond to unload logs), but for our purposes, they will be positioned so that they are just above the spar tree. Meanwhile, the locomotive takes back the empties and positions them on the outbound track (for empties) of the yard.

The other locomotive, having patiently waited for the boom crew to load the cars, would move up to the spar tree and pick up the loads for a trip to the mill. And the cycle then begins again.

The action down at the finished lumber shipping sidings is a more typical loose-car operational setup. You could have an old wheezing 4-4-0 or 2-6-0, long since removed from the main line, working as the mill switcher, shuffling loads

and empties. The switcher could take cars to the mainline interchange tracks, or a mainline locomotive could come in and do the job.

Conasauga Pine Timber Co.

Designed for HO scale, this 4 x 8-foot layout takes the essential elements of the Willamette layout and reduces it to fit on a sheet of plywood or foam board. It also moves the locale from the Pacific Northwest to northwestern Georgia. The 4 x 8-foot size is a great starting point for a small, manageable layout.

The key operation is the movement of strings of loaded or empty cars from the sidings behind the sawmill. Because of the interesting details of the mill, I have given it a central place on the layout with all of the track level for operational simplicity. By using two locomotives (geared or rod) to work the loads and empties (which can meet in the front passing siding), combined with a switcher in the finished lumber loading area, this layout could

provide an interesting evening of activity for two operators.

Although this is essentially a beginner's layout, I still recommend code 70 track, which has commercial turnouts available. It could also be built as a dual-gauge layout with the logging lines done in HOn3 or HOn2½ and the finished lumber lines in standard gauge, making them distinct from the rest of the logging operations. If you go with HOn2½, by sticking with 18" radius curves, you could run one of the Maine two-foot-gauge brass locomotives that have been on the market in the past decade. Mixing two gauges makes the layout more challenging, but still within the grasp of a dedicated beginner.

The Snoqualmie Lumber Co.

This 4 x 8-foot layout is designed to have lightweight but dramatic scenery made of extruded foam. The track plan is simple, but still offers operational opportunities.

The layout is a simple figure eight that rises over itself in the center. The yard tracks are for train storage, and the spurs on either side of the bridge offer a place to display equipment as well as provide operations.

By having three tracks in the central yard, two trains, operating one at a time, could pick up and set out log cars at the spurs.

This is a perfect trial layout for modelers wondering if logging is for them. A geared locomotive or two with DCC and sound, coupled with steep mountain scenery and lots of trees and ground cover, could make this layout a fun alternative to a more operations-intense layout. By varying the types of trees and style of topography, this layout could be in California, Arizona, the Pacific Northwest, or the Appalachians.

The Nescher & South Valley

This layout, originally published in the spring 1961 issue of *Model Trains* as the Nescher & South

The West Side Lumber Co. featured well-maintained track for a logging line, but note the varying tie lengths, light rail, and low ballast profile. *Bart Gregg*

Valley Mineral RR, is one of the best small layouts I've seen designed for sectional track. Don Reschenberg designed it as a coal hauler, but by changing the industries to represent pulpwood harvesting and loading, this layout could be a logging opera-

tor's delight. Locomotives could range from a Bachmann low-drivered 4-6-0 (or Richmond 4-4-0 model with DCC and sound) to GE 44- and 70-tonners or EMD or Alco switchers.

The equipment requirements are simple – one or two locomotives, a variety of 40-foot pulpwood flatcars, and some miscellaneous freight cars for the non-pulpwood industries.

Set in the rolling hills anywhere in the Southeast or upper Midwest, this layout gives transition-era modelers the opportunity to have steam and diesel on the same layout. Many of these small short lines kept a steam locomotive on the property for backup many years after the diesel arrived.

For those with a taste for sawmills, a small mill model such as Keystone's Danby mill, which receives its logs via truck, could also become a shipper of wood chips. This gives the option for more variety in freight cars. I would recommend locating the mill alongside the spur at the end of the Nescher passing loop, with the switchback tail rising up behind it.

Although a line like this should be a weed-grown, light-rail vision of backwoods short line, it was designed to be built with sectional track. You can do this, or substitute flextrack and turnouts with lighter rail.

Visiting a restored logging line and viewing the equipment can help you get a feel for modeling. This 1981 view shows West Side No. 15 on the West Side & Cherry Valley, which at the time operated on former West Side track.

Going beyond the obvious

The first six chapters of this book covered the basic information needed to start modeling a logging or forest products railroad. The focus was on what you could do with currently available equipment to create a scale model of a logging railroad. This chapter presents ways to model logging railroads that go further and includes the topics of fine scale modeling (intense, highly accurate detailing of equipment and structures), large- and small-scale modeling, and animation.

The fine scale alternative

The term "fine scale" has had many meanings over the years, and under current model railroad usage, it means using wheel and track standards scaled directly from the prototype.

Many logging scenes, like this one on the West Side Lumber Co., lend themselves to modeling in fine scale or large scale to capture the many intricate details. *Stan Kistler*

SEVEN

Restored West Side Lumber Co. Shay No. 15 rounds a curve with a log car in tow while it was operating on an Oregon tourist line in the 1970s.

Fine scale modeling has established a following among modelers in both HO (Proto:87) and O scales (Proto:48).

Early HO models of the 1940s and '50s were correct for their scale and gauge, but standards for wheels were coarse – wheel treads and flanges were significantly oversize to increase operational reliability. The National Model Railroad Association later established its RP-25 wheel standards, which greatly improved appearance but stopped well short of using prototypical dimensions.

Modelers' desire for something better led to the Proto:87 movement, which specifies prototype standards for both track and wheels. The move has been supported commercially by Northwest Short Line and other manufacturers. Ian Rice described the construction of a Proto:87 project layout in an eight-part series for *Model Railroader* (Octo-

ber 2003 through May 2004). You can find more information at www.Proto87.org.

Fine scale modelers in O scale follow similar practices, but also correct O's scale/gauge mismatch: A series of compromises early in the hobby's history left standard O scale track with a gauge of five feet instead of the prototypical 4'-8½". You can learn more about this area of the hobby at www.proto48.org.

On3 modelers have, almost by default, been working in fine scale all along. Grandt Line, Kemtron, Precision Scale, San Juan Models, and other manufacturers have made wheels to exact scale standards for many years. As a result, commercially available track, especially turnouts, follow exact or fine scale standards.

Many logging modelers have long built models and layouts to a higher level of detail. Some are interested in creating small but

intensely detailed layouts because of space limitations. Others do it because they enjoy focusing on the models themselves rather than on extensive operation.

The concept of the small but exact-scale layout with only one or two locomotives offers an opportunity for modelers who want to expand their skill base but not take on the challenge of a large layout without knowing if they can do it. It can also provide modelers a chance to try a different era than what they currently model.

Large scale possibilities

I'm an easy touch for geared steam locomotives, especially Shays. I own models of Shays in four scales and seven gauges, and would probably own even more if rational thought didn't stop me. One in particular that I have found very intriguing is Bachmann's two-truck Shay in large scale (1:20.3). Short of live steam in larger sizes, this model is the most fascinating I've found for its size and running capabilities. Although the wheels and flanges are not to scale, the size alone is enough to make me ponder the possibilities of building a true logging layout in this scale and gauge.

Many garden or outdoor layouts aren't built to scale, but that doesn't have to be the case. I'm not advocating putting real logs on those Bachmann skeleton cars, dumping them in a goldfish pond in the center of a layout, pulling them up on the saw chain, and actually cutting them in an animated sawmill. However, the possibility of a true walkaround, radio-controlled outdoor layout – one complete with steep grades that would make the Shay work hard as it climbs – is something even a devoted indoor railroader like me finds compelling.

A scale logging layout outdoors, with some compromises on gauge or wheel standards, constructed to what would be considered a normal indoor-layout height (in other words, close to eye level most of the time) is where my yearnings lie.

Taking a cue from the late and great John Allen, whose HO Gorre & Daphetid layout featured areas of floor-to-layout scenery, I would do the same thing outdoors. The walkways by the track could be cut into the earth, or dirt could be piled up high enough to make realistic mountains and valleys. Trees would have to be a compromise between what looks best from the local garden store and what will grow in your region, but a solid hill of evergreens with a few deciduous bushes for color in the fall might be the ticket for a true layout in the sun.

Imagine walking through the mountains as your Shay gingerly works its way down the hillside and over a pole trestle at a scale eight miles per hour, the soft

Scenes like this turn of logs in the Cascades in 1899 would make great focal points on a model railroad. *Darius Kinsey, Library of Congress*

chuffing of the sound system telling you the locomotive isn't working hard but the brakes probably are!

The same rules of track planning apply outdoors as indoors, but perhaps with fewer constraints, depending on your available space.

The live-steam experience
I have seen only one large-scale (12" gauge) live-steam geared locomotive – that is, one that is actually powered by steam – and it was a two-truck Heisler that was not working well that day. The boiler, which burned coal, was not firing well (shades of prototype reality!), but when it did run, the movement of the piston rods and central crankshaft was purposeful, and you could feel that this complex machinery was doing a real job, not just going along for the ride as with electrically powered models.

The amount of work needed to build a live steam logging locomotive in large scales (with track gauges from 7.25 through 15 inches) seems overwhelming. However, reference books listing parts suppliers for Shays and other geared locomotives are available.

Although this is far from scale model railroading in the basement or spare room, it would be a most interesting challenge. A live-steam Shay in 7.25-inch gauge wouldn't be used for switching on a giant layout, but would definitely give the operator a better sense of what it was like to operate a real Shay than one can get from an electrically powered model.

For the person for whom a Proto:48 or 1:20.3 Shay just isn't big enough, this may be the ticket to future happiness.

Potential downsides include cost and time. Live steam is a hobby unto itself, and requires having or acquiring machining and other skills.

Animation
Just as we need to learn to think of track as a model, perhaps we also need to think of the movement of the timber to the loading area as being as much a part of logging railroad operation as the movement of the trains.

For some model railroaders, the mention of animation calls to mind many of the highly unrealistic three-rail and toy-train accessories of the 1950s and '60s. However, animation can be applied in more realistic fashion.

An outdoor layout could be built using this West Side Lumber Co. scene as inspiration. *Stan Kistler*

One of the limitations of a logging railroad is that the most interesting and important parts of the activity, the loading of the cut timber onto log cars and the subsequent dumping of those logs into the mill pond or cold deck, seem to be impossible to re-create in scale. But can some of the electronic power of Digital Command Control technology be applied to other areas of the layout that are part of the operation of a logging railroad?

Digital Command Control has given us the ability to have working Mars lights on our diesels, a form of animation that works well. This moving headlight beam strikes a chord in any railfan who has ever watched a train at night. And sound, another form of animation for both steam and diesel, is a reality for almost any DCC-equipped locomotive. The ability to turn lights on and off

from a DCC controller is animation, as is the ability to remotely actuate switch motors.

The ability of DCC systems to control motor speeds in up to 128 steps provides the possibility of having movement that starts slowly, increases in speed as it would in real life, and then gently comes to a halt. For example, with proper design of the motor and moving mechanism, one could build a motorized, DCC-controlled Barnhart loader or steam skidder. Running the loader could be a job during an operating session.

Heljan recently released a working DCC-controlled container crane model in HO scale, showing that such devices are possible. The item may also offer possibilities for conversion into log-loading duties.

Another potential project would be modeling the movement of logs over a high-lead system, with a

mechanism to drop logs onto the hot deck by the loading boom. This would be similar to the loader, but requiring special attention to the construction of the cables.

The next step would be for an enterprising modeler to develop a system for unloading logs at the mill. Again, DCC provides potential tools for doing it.

Looking for inspiration

I encourage to anyone interested in the logging industry and logging railroads to check one out. There are museums across the country that have geared steam locomotives, operating equipment, and displays that cover nearly every aspect of logging. It's hard to describe in words the actions and sounds of a Shay or Heisler moving under steam, and even video is a poor substitute for the real thing.

The Cass Scenic Railway, in the mountains of West Virginia, is probably one of the best on the East Coast; the Mt. Rainier line in western Washington is an excellent one on the West Coast. Both showcase operational logging locomotives. There are many other good railroads, and checking the annual *Trains* magazine guide to recreational railroading or the *Tourist Trains Guidebook* (published by Kalmbach) is a great way to start.

At local historical societies and libraries, you can search for old newspaper articles, written histories of an area, and photograph collections. Perusing county records for businesses long gone, articles of incorporation, and court records for accidents and contracts can also provide helpful information.

For example, I live near Lumber Company Road, which is a dead giveaway that something timber-related happened nearby. Conversations with older residents led me to the spot (now covered with trees and underbrush) where a sawmill used to stand. The remnants of a long-unused siding are still in the weeds, rails intact, but with no connection to the main line. They stand as a silent witness to carloads of pine lumber that once were shipped over the old Hook and Eye route of the Louisville & Nashville. I also learned that at one time several narrow-gauge logging operations in my county provided not only pine logs for sawmills but props and timbers for the local marble industry. There is more, but I haven't researched it all yet. It is this sort of journey that makes modeling a logging line more than just another layout-building exercise.

Group inspiration
The Internet has given us a wonderful resource for modeling. Web pages cover nearly every aspect of model and prototype

Remember that logging was an industry, not a pastime. Logging railroads were about the logger and the logs more than the railroad. *Photo courtesy Weyerhaeuser Archives*

railroading, and there are discussion groups for every interest, scale, and gauge. Joining a group that suits your interest, or even starting one if there isn't one that works for you, is a way to connect with a community of like-minded people who will inspire you to model more and procrastinate less. The ability to post photographs on group sites has driven me to finish more half-completed models than I would have ever back in the "dark ages." Back then, it took an open house or show to inspire me to complete a

model. Now, I get pinged by the monthly "what's on your workbench" question, which pushes me to do something each month.

The Internet is a great place to find information, but it doesn't replace books or monthly magazines for permanence and reference standards. A multitude of books are available, focusing on both the logging industry as well as the railroads themselves. And www.kalmbach.com has links that can take you to almost anywhere in the model railroading universe. I hope you enjoy the ride!